WHAT'S KILLING
MY CHICKENS?

WHAT'S KILLING MY CHICKENS?

THE POULTRY PREDATOR DETECTIVE MANUAL
Identification • Protection • Prevention

GAIL DAMEROW

Storey Publishing

The mission of Storey Publishing is to serve our customers by
publishing practical information that encourages
personal independence in harmony with the environment.

Edited by Deborah Burns
Book design by Erin Dawson and Michaela Jebb
Indexed by Samantha Miller

Cover photography by © Anthony Lee/Getty Images, front, b.; © Mark Chrimes/EyeEm/Getty Images, back, 2nd from t.; © Melissa Goodwin/ Getty Images, back, b.; © Nature Picture Library/Alamy Stock Photo, back, 2nd from b.; © schankz/stock.adobe.com, back, t.; © Tambako the Jaguar/Getty Images, front, t.l.; © Todd Ryburn Photography/Getty Images, front, t.r.

Interior photography credits
© AGAMI Photo Agency/Alamy Stock Photo, 149 r.; © Alexander McClearn/ Alamy Stock Photo, 6 t.; © alexandrumagurean/iStock.com, 200; © All Canada Photos/Alamy Stock Photo, 178, 186, 216; © ands456/iStock.com, 22; © ArendTrent/iStock.com, 139 l., 257 r.; © Art Directors & TRIP/Alamy Stock Photo, 172 l.; © Arterra Picture Library/Alamy Stock Photo, 148; © Avalon/ Photoshot License/Alamy Stock Photo, 164 b.; © AYImages/iStock.com, 174; © BGSmith/stock.adobe.com, 240 r.; © Bill Gorum/Alamy Stock Photo, 245 b.; © BirdImages/iStock.com, 234; © blickwinkel/Alamy Stock Photo, 230 r.; © Bob Hilscher/Alamy Stock Photo, 164 t.; © bookguy/iStock.com, 143 r.; © Brian E Kushner/iStock.com, 131, /stock.adobe.com, 242; © ca2hill/iStock .com, 149 l.; Carl D. Howe/Wikimedia Commons, 251; © Christina Krutz/ Getty Images, 6, 2nd from t.; © cpaulfell/iStock.com, 223 r.; Dakota L./ Wikimedia Commons, 254 l.; © David Chapman/Alamy Stock Photo, 2 t.r., 136 row 4 r.; © Denja1/iStock.com, 191; © Dennis Holcomb, Dennis Holcomb Photography/iStock.com, 152; © Design Pics Inc/Alamy Stock Photo, 6 b.; © Dgwildlife/iStock.com, 171 r.; © donyanedomam/stock.adobe.com, 189; © Dunca Daniel Mihai/Alamy Stock Photo, 175; © Dvorakova Veronika/stock .adobe.com, 54; © fotoguy22/iStock.com, 197; © FRANKHILDEBRAND/ iStock.com, 144; © Gail Damerow, 15, 32, 46, 91, 100, 103; © Gavin Thorn/ Alamy Stock Photo, 2 b.r.; © geoffkuchera/stock.adobe.com, 221 l.; © gsagi/ iStock.com, 2 b.l.; © Harry Collins/Alamy Stock Photo, 172 r.; © Henry W. Art, 207; © hkuchera/stock.adobe.com, 89; Howcheng/Wikimedia Commons, 225; © hstiver/iStock.com, 138 r.; © imageBROKER/Alamy Stock Photo, 170 l., 182; © inga spence/Alamy Stock Photo, 6, 3rd from t.; © Jason Ondreicka/Alamy Stock Photo, 244, 250; © jaypetersen/iStock.com, 145; © jcrader/iStock.com, 212; © Jganz/iStock.com, 136 row 3 r.; © Jim Merli/ Getty Images, 2 t.l.; © Joe Amon/Getty Images, 130; © Joseph Kirsch/stock .adobe.com, 118 r.; © jsbb123/iStock.com, 156 l.; © KeithSzafranski/iStock .com, 150 r.; © Kneonlight/iStock.com, 150 l.; © Krys Bailey/Alamy Stock Photo, 183, 185; © kwasny221/stock.adobe.com, 209; © Linda Freshwaters Arndt/Alamy Stock Photo, 187; © lizaveta25/stock.adobe.com, 126; © Lynn_ Bystrom/iStock.com, 199, 202; © Mark Chrimes/EyeEm/Getty Images, 5; © Mark Kostich/iStock.com, 248; Mars Vilaubi, 122; © Matt Dirksen/iStock .com, 165; © Melissa Goodwin/Getty Images, 85; © Michelle Gilders/Alamy Stock Photo, 71, 254 r.; © mikespics/iStock.com, 157; © Mircea Costina/ Alamy Stock Photo, 136 row 1 l.; © mtruchon/stock.adobe.com, 214; © NCBateman1/stock.adobe.com, 30; © NNehring/iStock.com, 136 row 2 l.; © NPS Photo/Alamy Stock Photo, 171 l.; © Paolo-Manzi/iStock.com, 141; © Paul Hawkett/Alamy Stock Photo, 83; © Paulo Oliveira/Alamy Stock Photo, 239; © pchoui/iStock.com, 156 r.; © phototrip.cz/stock.adobe.com, 140 l.; © Prisma by Dukas Presseagentur GmbH/Alamy Stock Photo, 245 t.; © Richard Hayward/Alamy Stock Photo, 188; © Richard Mittleman/Gon2Foto/ Alamy Stock Photo, 136 row 3 l.; © Rick & Nora Bowers/Alamy Stock Photo, 136 row 4 l., 222 r., 223 l.; © Robert Shantz/Alamy Stock Photo, 140 r., 170 r., 240 l.; © robin chittenden/Alamy Stock Photo, 136 row 2 r.; © Rohan/stock .adobe.com, 118 l.; © RussellGr/iStock.com, 257 l.; © sandf320/iStock.com, 198; © Shravan Sundaram Photography/iStock.com, 138 l.; © SteveByland/ iStock.com, 35, 136 row 1 r.; © subjug/iStock.com, page 6 and throughout (background); © Susan E. Degginger/Alamy Stock Photo, 137; © Tambako the Jaguar/Getty Images, 10; © Terry Sohl/Alamy Stock Photo, 139 r.; © twildlife/ iStock.com, 193; © vazaluza /stock.adobe.com, 128; © Vince Burton/Alamy Stock Photo, 146; © vsanderson/iStock.com, 143 l.; © Whiteway/iStock.com, 230 l.; © win247/iStock.com, 160; © Zoran Kolundzija/iStock.com, 33; © ZUMA Press, Inc./Alamy Stock Photo, 232

Illustrations by © Elayne Sears
Maps, silhouettes, tracks, and graphics by Ilona Sherratt

Storey Publishing
210 MASS MoCA Way
North Adams, MA 01247
storey.com

Printed in China through World Print
10 9 8 7 6 5 4 3 2 1

LIBRARY OF CONGRESS CATALOGING-IN-PUBLICATION DATA

Names: Damerow, Gail, author.
Title: What's killing my chickens? : the poultry predator detective manual / by Gail Damerow.
Description: North Adams, MA : Storey Publishing, [2019] | Includes bibliographical references and index.
Identifiers: LCCN 2019015503 (print) | LCCN 2019019186 (ebook) | ISBN 9781612129105 (Ebook) | ISBN 9781612129099 (paperback : alk. paper)
Subjects: LCSH: Chickens. | Chickens—Predators of.
Classification: LCC SF487 (ebook) | LCC SF487 .D187 2019 (print) | DDC 636.5—dc23
LC record available at https://lccn.loc.gov/2019015503

Storey books are available at special discounts when purchased in bulk for premiums and sales promotions as well as for fund-raising or educational use. Special editions or book excerpts can also be created to specification. For details, please call 800-827-8673, or send an email to sales@storey.com.

Acknowledgments

Many people deserve thanks for their hand in helping develop this book. Among them are those who took time from their busy lives to fill gaps in my knowledge: Rebecca Waters, the Wolverine Foundation; Jackie Marsden, Squirrel Refuge; Dr. Eric Yensen, professor emeritus, College of Idaho; Dr. Gail R. Michener, professor emeritus, University of Lethbridge; Jeff Beane, herpetologist, North Carolina Museum of Natural Sciences; Dr. James O. Farlow, crocodilian expert, Purdue University; and the ever patient Dr. Eugene S. Morton, professor emeritus, York University, who patiently responds to my constant barrage of bizarre bird questions.

Thanks go also to the poultry keepers who provided firsthand accounts of their own experiences with poultry predation: the late Bethany Caskey, my favorite illustrator; the Chicken Chick, Kathy Shea Mormino; Victoria Redhed Miller, author, *Pure Poultry*; Dave Holderread, Holderread Waterfowl Farm; James Stockton, Stockton Farms; Paul K. Boutiette, founder, eggcartons.com; Pamela Art, retired publisher/president, Storey Publishing; Diana Mitchell, Los Angeles Urban Chicken Enthusiast; Celeste Tittle, Ham & Eggs Ranch; and Frieda McArdell, backyard pond builder extraordinaire.

A big thank-you goes to all the folks at Storey Publishing, particularly editor Deb Burns, who shepherds my manuscripts through the publishing process, and photo editor Mars Vilaubi, who tolerantly tutors me on how to take decent photographs.

And finally, thanks to my forbearing husband, Allan, who surely looks forward to enjoying dinner conversations that do not involve poultry predators.

CONTENTS

On Becoming a Detective 8

PART ONE: THE OFFENSE 10

PART TWO: THE SUSPECTS 130

On Becoming a Detective

My first flock of chickens came with a 1-acre (0.4 ha) ranchette in a rapidly suburbanizing area, where the chief threats were dogs, rats, and ever-tightening zoning regulations. For the past 35 years, however, my husband, Allan, and I have happily raised chickens and other poultry on a densely wooded Tennessee farm at the end of a gravel lane, where we feel privileged to occasionally spot a bobcat, coyote, raccoon, fox, skunk, or opossum (and, once, a bear!) wandering through our farm. Much of the time we wouldn't be aware of these potential predators passing within yards of our poultry yard — demonstrating no interest in our various flocks — if it weren't for the motion-sensitive cameras we use to monitor their comings and goings.

Shortly after we moved to this farm, the neighbors informed us that it is impossible to keep poultry here; sooner or later any flock we attempt to keep would fall victim to one of the many predators that lurk nearby. Undeterred, we immediately raised a flock of laying hens. Soon, ducks and geese were swimming in the pond behind our house, turkeys were parading around our barn, and guinea fowl were patrolling our orchard.

During the decades we've lived here, our various flocks have, for the most part, coexisted with the wildlife. The operative phrase is "for the most part," because every now and then one of those magnificent creatures that share our farm visits our poultry yard, resulting in an unhappy outcome for one or more of our domestic birds.

Allan and I console ourselves by reminding each other that wild animals have to eat, too. Just as we plant extra vegetables in our garden to have enough for ourselves after the rabbits, rodents, and songbirds take their share, we generally raise more chickens and other poultry than necessary to satisfy our personal need for eggs and meat.

But still, it's maddening to have a bobcat make off with one of our turkey hens. And it's difficult not to get upset when an owl kills a favorite rooster and leaves all but the head. What a waste.

Once we get over the initial shock of losing a bird, we set about trying to identify the perpetrator. Then we look for ways to deter it by improving our poultry yard security. Maybe the electric fence isn't working as well as it should be. Maybe our solar predator lights are no longer recharging. Maybe the chickens, or the predators, have scratched out a gap under a gate. Maybe we need to keep our birds confined indoors until later in the morning.

The way we see it, the wildlife was here first. We, along with our domestic poultry, are the intruders. Because we introduced poultry into this forest, it's our job to ensure that our birds are not more enticing and easier to obtain than the predators' normal and natural fare.

Too many poultry keepers get preoccupied with frustration killing. When I run across an online forum describing an attack by one or another predator, the discussion typically focuses on ways to kill or relocate the unfortunate creature. What a sad state of affairs when poultry keepers are more interested in the legalities of eliminating a predator than in learning better ways to protect their flocks.

The premise of this book is that a predator attack is the fault not of the predator, but of the poultry keeper. Those of us who undertake poultry keeping take upon ourselves the obligation to keep our birds safe from harm.

This book is, therefore, not about how to trap or kill predators that attack your poultry. Rather, it explains why killing or relocating predators is a decidedly bad idea that, ultimately, is only a temporary measure. This book will help you determine which predators are likely to appear in your area (and, just as important, which are not), give you insights into their behavior, and use the information to devise effective ways to keep your poultry safe. Along the way, I hope you will come to appreciate and enjoy the wildlife that share your little patch of earth.

How to Use This Predator Detective Guide

This book is organized into two parts. The first six chapters offer an overview of poultry predation, discussing the nature of predators, how to identify a predator that has taken an interest in your poultry, and steps you can take to protect your birds using practical, nonlethal solutions.

The remaining chapters profile individual predator species, describing such important details as their telltale signs, behavior patterns, and documented geographic range. After you use the guidelines in part 1 to determine what type of predator you most likely are dealing with, part 2 will help you narrow your choices based on where you live and the clues you find. Deterrent options suggested in these profiles are described more fully in chapters 4 (Controlling Predators) and 5 (Fence Defenses).

The appendix includes a predator worksheet. Consult the maps included with the predator profiles and use this worksheet to mark the predators that are likely to appear in your area. Realize, however, that individual animals can stray quite far from the known population range for their species. When you are faced with a predator problem, this worksheet will help you quickly focus on probable suspects.

Every poultry keeper experiences predation at one time or another. A variety of wildlife, especially those feeding young, take an interest in chickens and other poultry as quick and easy meals. The more we humans encroach on their habitat, the smarter these animals get, and the smarter they get, the smarter you have to be to outsmart them.

This book is here to show you the way.

Gail Dameron

PART ONE
THE OFFENSE

1. SCENE OF THE CRIME

"There is nothing like firsthand evidence."

Sherlock Holmes

Everyone who keeps poultry sooner or later experiences that heart-stopping moment of realization that a predator has come to call. A typical first reaction is to rush in, assess the damage, and clean up the mess. But stopping to carefully survey the scene can give you valuable clues toward determining what type of predator was involved, and therefore what precautions you can take to prevent a future recurrence.

Although this initial survey can be extremely helpful in narrowing down the list of potential predators, it may not provide a conclusive identification. Further clues may be found in the form of tracks, scat, and other signs left at the scene, as described in the next chapter. In an active poultry yard, however, where such signs already may have been obliterated, your first best guide is to examine where, how, and when birds died or went missing.

POULTRY PREDATION

Each predator species has a typical way of killing and consuming its prey. "Typical" is the operative word. Not all animals within a single species work in exactly the same way. Further, younger members of the species may work in a slightly different manner than the older, more experienced members. Still, certain clues can point you in the right direction. For starters, mammals and raptors leave different sets of signs.

When birds are missing from the poultry yard, among the important initial clues to note are how many are missing, their age and size, and any damage to the fence or coop that may indicate whether the predator walked or flew. When birds are left dead at the scene, again note the number of birds involved, their age and size, and the condition of the facilities, as well as the appearance of the remains. Photographing the remains will give you a visual record of which parts were left behind, as well as recording the day and time of your discovery.

Chicks Missing

When baby poultry go missing without a trace, the culprit is usually a snake, a rat, or a house cat. None of these predators is capable of carrying off a mature chicken but can disappear one or more chicks in no time flat. In general, when small birds are missing, suspect a smaller predator. Larger predators capable of carrying a mature chicken, or even a turkey or goose, are more likely to do so than bother with smaller, younger ones.

A snake will eat whole chicks, leaving no evidence behind, except maybe the snake itself. We once found a black rat snake in our homemade wood-and-wire brooder after it had gulped down a couple of chicks and then — having gotten too fat to slip back out through the wire — curled up under the heat lamp to sleep off the fine meal.

Rats, too, can disappear baby chicks without a trace. A rat will pull a baby chick down into its tunnel, but if the bird is partially grown it may get stuck at the entry. You may find the bird, having been pulled head first, with its feet sticking out of the tunnel opening.

Domestic and feral house cats can easily disappear baby birds. One year, I lost a batch of chicks to a feral cat that was feeding her kittens in a nearby vacant lot. Another year, I lost several goslings that were housed with their parents behind a chain-link fence. I couldn't fathom what was getting to them until I happened to see the little goslings pop through the fence to graze on the lawn, then have trouble squeezing their fattened bellies back through the fence. Nearby, intently watching the goslings, was the neighbor's cat.

I accidentally learned the best way to train a cat to leave chickens alone when my own new kitten followed me to the chicken yard. She took an interest in some baby chicks following a mother hen, whereupon the hen puffed up to twice her normal size and charged the kitten. For the rest of her life, that cat laid her ears back and skulked away whenever a chicken got close.

Other potential threats to baby poultry include ground squirrels, foxes, skunks, minks, and weasels. Among predatory birds, those that favor baby poultry include the smaller hawks — such as sharp-shinned and zone-tailed hawks — along with crows and ravens. Ducklings and goslings on open water are additionally susceptible to being nabbed by a snapping turtle, a large fish such as northern pike or largemouth bass, or a young alligator.

DON'T DISCOUNT HERBIVORES

Numerous accounts have documented cattle, sheep, deer, and other normally herbivorous animals eating baby poultry or even eggs. Some researchers believe that once an herbivore accidentally gets a taste of chick or egg while grazing, it will deliberately seek more of the same. Others speculate that the unusual appetite may be triggered by a mineral deficiency, most notably calcium. In any case, if baby poultry or eggs go missing where cattle, sheep, or deer graze, don't overlook the possibility that an herbivore may have devoured them.

Grown Birds Missing

Mature poultry that disappear without a trace may have been carried off by a fox, coyote, dog, bobcat, eagle, owl, or hawk. Foxes and coyotes can disappear a large number of birds within a short time without leaving any signs, and they will hunt during the day when feeding young. A bobcat usually takes one bird at a time and may leave a trail of feathers. Domestic dogs are particularly careless about scattering feathers, along with injured and killed poultry.

Although a hawk rarely carries off a large chicken, a big hawk can snag a bantam, as well as a growing chicken or other type of poultry. An owl, too, may carry away a small bird, but it is more likely to leave a dead bird with just the head missing. Neither hawks nor owls are shy about marching right into the coop for a snack. One morning I opened the coop door to find a young owl snugged in among the roosting chickens. Another time I found a hawk inside the coop, wrestling with a sizable cockerel.

An eagle usually carries the bird away from the coop and then picks its bones clean, leaving nothing but a skeleton, sometimes not far outside the poultry yard. After missing one of our chickens, we once found this calling card in the middle of our hay field.

Raccoons, too, will carry off a chicken or duck. They may raid the coop as a cooperative venture and then squabble over their kill. As with an eagle, you may find the carcass some distance from the house, but only the organs will be eaten and the feathers will be scattered around.

On rare occasions, a house cat (usually feral) will kill a mature chicken, eating the meatier parts and leaving the skin and feathers, and sometimes other parts, scattered around. A bobcat usually carries away the bird to eat elsewhere and may bury, or cache, any uneaten part to snack on later. A fox or mountain lion will do the same. You can sometimes find the cache by following a trail of feathers and blood. Coyotes and bears also cache; in their case the carcass is more likely to be torn apart rather than left largely in one piece.

Birds Dead or Wounded

Bitten birds, either dead or wounded, may have been attacked by a dog. If they are young birds and the bites are around the hock, suspect a rat, which may also seek protein by pulling off and eating feathers from birds on the roost. If the bites are on the leg or breast, the biter is likely an opossum. Possums like tender growing birds and will sneak up to the roost and bite a chunk out of the breast or thigh of a sleeping bird. On the rare occasion that a possum outright kills a chicken, it usually snacks right then and there.

Chickens found dead in the yard, but without any missing parts, were probably attacked by a dog. When the bird stops moving, the dog loses interest, leaves the bird where it died, and seeks a livelier

ASSESSING PUNCTURES

Examining a dead bird for punctures left by teeth or claws can give you helpful clues. The easiest way to assess puncture marks is to skin the victim and examine both the underside of the skin and the flesh beneath the wounds in the skin. Punctures made by a mammal's teeth are typically not as deep as those made by a raptor's talons. The tissue between a pair of tooth punctures is typically crushed, but rarely so if the wounds were made by talons.

Poultry with talon punctures may also have beak punctures on the back and breast. If the victim's breast has been eaten away, look for triangular beak marks along the breastbone.

If, on the other hand, you believe the wounds were made by teeth, their size and distance apart can provide additional clues in identifying the predator responsible. For example, tooth punctures made by a coyote will be larger, deeper, and at least an inch apart compared to the smaller, shallower, and more closely spaced marks left by a weasel. After measuring the distance between puncture marks, consult the table Average Distance between Canine Teeth Punctures on page 261.

When threatened with bodily harm, guinea fowl typically release handsful of undamaged feathers, leaving you to seek other clues.

playmate. Don't discount the possibility that the perp might be your own dog. I once found a dozen dead fryers lined up neatly on the walkway next to my fish pond. I was gazing at them while trying to guess what kind of predator could have done such a thing when my new puppy came happily bounding up with yet another bird to add to his collection.

Like dogs, weasels and their relations (ferrets, fishers, martens, minks, and so forth) engage in killing sprees. If you find bloodied bodies surrounded by scattered feathers, you were likely visited by a member of the weasel family. Fishers and martens may kill and stash chickens to come back and eat later. Weasels can sneak into housing through an opening as small as 1 inch. They sometimes run in family packs that can do significant damage in an amazingly short time.

If you suspect a predator has been casing your coop, and you find dead birds with no obvious injuries other than possibly being flattened, the only thing you know is that the predator frightened them. In trying to get away, they piled in a corner or against a wall and the birds unfortunate enough to end up at the bottom of the pile suffocated.

Parts Missing

Parts missing from a dead bird can help you identify the culprit. A chicken found next to a fence or in a pen with its head missing is most likely the victim of a raccoon that reached in, grabbed the bird, and pulled its head through the wire. Or a bird of prey could have frightened your chickens into fluttering against the wire, and any that got stuck in the wire had their heads bitten off.

When you find a bird dead inside an enclosure with its head and crop missing, your visitor was a raccoon. If the head and back of the neck are missing, suspect a weasel or mink. If the head is missing and the front of the neck is eaten, perhaps down into the breast, and your bird has been plucked, the perpetrator is a raptor.

Just as a raccoon will reach into a pen and pull off a chicken's head, so will it also pull off a leg, if that's what it gets hold of first. Dogs, too, may prowl underneath a raised pen, biting at protruding feet and pulling off legs. Birds bitten around the rear end with their intestines pulled out have been attacked by a weasel or one of its relatives.

SCENE OF THE CRIME

Poultry Predator Detective Key

ONE OR TWO BIRDS KILLED

TYPICAL CLUES	POSSIBLE PREDATOR	TYPICAL CLUES	POSSIBLE PREDATOR
Head bitten off, claw marks on neck, back, and sides; carcass carried away and partially covered with litter; prefer larger poultry (turkey or goose)	bobcat,* cougar, lynx	Chicks pulled into fence, wings and feet not eaten; feathers and parts of young birds strewn, meaty portions eaten, tooth marks on bones, wings and loose skin with feathers remain	cat (domestic or feral)
Skeleton picked clean, scattered feathers	coyote	Breast and legs eaten, small bones bitten through, toes curled, remains scattered or buried	fox
Young or small birds bitten into	ground squirrel	Only head and front of neck (or part of breast) missing; or entire chicken eaten on site, only skeleton and plucked feathers remain	hawk
Entrails eaten through cloaca; breast or thigh of young bird bitten into (bird possibly still alive)	opossum	Only head and front of neck (or part of breast) missing, or deep marks on head and neck; maybe feathers scattered around fence post	owl
Only head missing; or neck, breast, and crop chewed, entrails may be missing; bird pulled into fence or pen wire and partially eaten; maybe carcass found away from housing, with feathers scattered	raccoon	Bruises and bites on legs of young birds; or partially eaten with head down tunnel	rat
Chicks killed, abdomen torn out or eaten (but not muscles and skin), maybe lingering odor; older birds mauled, neck opened up, heads missing	skunk	Large poultry (goose or turkey) disemboweled	wolf

MULTIPLE BIRDS KILLED

TYPICAL CLUES	POSSIBLE PREDATOR	TYPICAL CLUES	POSSIBLE PREDATOR
Birds torn apart or missing, intestines scattered, fence or building torn into, scat nearby	bear	Some birds missing, some remain with broken necks, feathers strewn around and trailing away	coyote,* fox
Birds mauled and bitten all over body, with broken necks, but not eaten (one may be missing); fence or building torn into; feet pulled through cage bottom and bitten off	dog	Rear end bitten, intestines pulled out; not much eaten (may cache birds to eat later); feathers widely scattered	fisher, marten
Hatchlings with tiny tooth punctures in head; bodies of small birds neatly piled, killed by small bites on head, neck, and body; back of head and neck missing; possible musky odor	mink	Bites on neck and body, bruises on head and under wings, back of head and neck missing, intestines pulled through cloaca, bodies neatly piled; possible faint skunk-like odor	weasel
Chicks dead, possible faint lingering odor	skunk	Heads and crops eaten, sometimes also breast; gate may be open	raccoon

TYPICAL CLUES	POSSIBLE PREDATOR	ONE BIRD MISSING
No clues, or feathers trailing near water	alligator, river otter	
No clues, or a few scattered feathers	hawk, owl	
No clues, or feathers scattered or trailing	bobcat, coyote, fox, cougar, lynx	
Fence or building torn into, feathers scattered	dog	
Small bird missing, no clues or a few scattered feathers, possible lingering musky odor	mink	

TYPICAL CLUES	POSSIBLE PREDATOR	MULTIPLE BIRDS MISSING
No clues, or feathers of large poultry (turkey or goose) scattered or trailing	bobcat	
No clues, young waterfowl missing	bullfrog, snapping turtle	
No clues or scattered feathers; possibly dead birds left behind with broken necks or neck feathers missing	coyote, fox	
No clues, or gates/doors open	human	
Chicks or young birds missing, wet feathers scattered	opossum	
Small birds missing, bits of coarse fur at point of entry, gates/doors open	raccoon	
Chicks or young birds missing, no clues	snake,* cat,* rat, raven	

*Most common

Note: Cross-reference the crime-scene clues detailed in this table with the timing of the attack, as best you can determine it. See the table on page 41 for details on the periods of greatest activity for the likeliest suspects.

Poultry Feathers as Clues

Poultry feathers left by a predator can provide important clues as to the animal's identity. If feathers are scattered around at random, the predator is likely a mammal. If feathers create a trail leading away from the poultry yard, the mammal is probably a bobcat or a fox. If feathers are concentrated within a circle, the predator is an eagle, falcon, or hawk.

Let's say, for instance, that you find the skeletal remains of a bird surrounded by a mess of feathers. Coyotes and raptors both eat the flesh off the bone and leave the skeleton largely intact. The pattern of the feathers, whether scattered or forming a circle, will tell you (respectively) whether you're dealing with a coyote or a raptor.

Damage to the feathers themselves provides another clue. Although it's not entirely definitive, feather damage can be helpful when combined with other clues found at the scene. Here's what to look for.

Quill shattered. A great horned owl uses its strong beak to pluck out feathers by grasping the base of the quill, leaving feathers whole and possibly with the quill end shattered.

Quill crushed. A falcon also uses its beak to pluck feathers by the base of the quill, leaving feathers whole with the quill end flattened.

V-shaped marks on shaft. A hawk plucks out feathers by grasping them partway up the shaft, often leaving V-shaped beak marks on the shaft.

Quills cut raggedly. A raptor may shear off feathers, leaving the tips of quills in its prey. The quill ends will likely not be flattened and will not show bite marks.

Feathers crimped. Besides plucking feathers with its beak, a raptor may also use its feet, leaving feather shafts bent in two places, like an elongated Z.

Quills cut neatly. A coyote may pluck out whole feathers but also uses its molars to bite off feathers where they enter the skin, neatly shearing off the tips of quills at a consistent angle and leaving them slightly flattened. Groups of feathers bitten off together remain as clumps, held together by saliva. A fox also shears feathers at the quill tip.

Damaged webbing. A fox uses its canine teeth to shear off feathers, usually farther up the quill than a coyote would bite, and may also pluck feathers (commonly tail feathers) by gripping them partway up the shaft, leaving the web rumpled looking.

Quills chewed. A bobcat uses its molars to chew off small clumps of feathers, leaving quills looking jagged rather than neatly sliced through.

Pieces of feathers. A weasel, with its small mouth and teeth, bites off feathers in pieces and at varying angles.

Flesh clinging to quills. Bits of skin clinging to the tips of feather quills indicate the feathers were plucked by a scavenger from an already dead bird. Clean quills indicate the bird was plucked immediately after being killed by a predator.

Disrupted Setting Hen

Having a nest raided of its eggs is bad enough, but when the nest includes a setting hen, the hen as well as the eggs could be lost. Of course you'd want to protect a setting hen for the duration of the brooding period, but sometimes a hen — particularly a duck, goose, guinea, or turkey hen — has ideas of her own. When a nest is raided, a few clues may help you determine whodunit.

Some egg predators — including badgers, crows, ground squirrels, raccoons, and skunks — will typically eat a hen's eggs without killing the setting hen. They may attempt to steal eggs out from under the hen, or they may harass her until she leaves the nest while they enjoy their meal.

A setting hen that voluntarily leaves the nest to eliminate, grab a bite to eat, get a drink, and (in the case of waterfowl) maybe take a quick swim covers her eggs with feathers and other nesting material before departing. If a predator startles her off the nest or has carried her away, the eggs will be left uncovered.

Some predators will eat both the hen and her eggs. Signs that a predator got the hen may be blood spots in or near the nest, and maybe a few random feathers. The dead hen may be found nearby, whole or partially eaten. If the latter, note whether the head is attached, detached, or missing.

A red fox will kill a hen by biting her in the neck and then will typically carry the hen away to eat elsewhere, leaving only a few feathers and perhaps a flattened area of vegetation where the hen was temporarily set aside while the fox filled up on eggs. A mink will kill a hen by biting her in the head or upper neck and then, after eating a few eggs, may drag the hen to its den for dinner. A coyote will partially eat the hen, often including the head, leaving the remains near the nest. A bobcat will typically snatch the hen and immediately carry her away, leaving you to wonder where she went.

EGG PREDATION

Many predators like poultry eggs as much as poultry keepers do. If and when such predators have access to your hens' nests will depend on the nature of your setup and on the type of poultry you keep. Indoor nest boxes are more likely to be visited by small, agile creatures. Outdoor nests — such as those favored by ducks, geese, guineas, and turkeys — are exposed to all the same predators plus the larger ones that tend to be wary of entering an enclosure from which escape may be difficult. Accordingly, your indoor nests are more likely to be visited by a crow, jay, ground squirrel, mink, or rat than by a badger, bobcat, coyote, or fox.

One of the first things to consider when eggs start disappearing is whether or not one (or more) of your own hens is eating them. Chickens, both hens and roosters, can develop the bad habit of eating eggs out of the nest. You may find an empty, partially eaten eggshell, or no shell at all. One sign of egg cannibalism is nesting material that's been dampencd by sticky egg goo. The egg eater will quite likely have a yolk-smeared beak. Another indication that a chicken might be the culprit is finding the same signs in several nests. Most other egg predators typically won't stay long enough to travel from nest to nest inside a coop.

Another consideration is the possibility that a nest full of empty shells was not the victim of foul play, but rather the eggs were incubated and hatched behind your back. This occurrence is more likely when the eggs have been hidden outside the coop proper. A hatching chick of any poultry species breaks out of its egg by rotating its head around the blunt end of the shell, chipping away a narrow slit until the top (approximately one-third) of the shell separates from the bottom two-thirds. The shell membrane typically curls inward around the edges of the opening, and the inside of the empty egg shows signs of blood vessels. Sometimes the smaller end piece gets turned around and becomes cupped inside the larger piece of shell.

An egg that was eaten by a skunk may initially look like it hatched, but if you examine the shell you'll see a difference. After making a hole in the end of the shell, the skunk licks out the contents, in the process pressing its nose against the edges of the hole and crushing them inward. Further, a skunk typically opens an egg at the pointed end, while a chick hatches at the blunt end of the shell. Lingering odor may or may not be an additional clue, since other egg predators sometimes also smell "skunky."

Once you have determined that the missing eggs neither hatched naturally nor were eaten by your own chickens, signs found at the scene can help you determine the predator's identity. Such signs include eggs missing without a trace, the appearance of any shells left behind, and disturbances in and around the nest.

Eggs Missing

Some predators carry eggs away from the nest without leaving any signs. That, itself, can provide a possible clue. When I find occasional eggs missing from nest boxes inside the coop, the first predator I suspect is a rat snake, especially if I've seen one hanging around hunting for rodents. Sometimes the snake is easy to ID because it curls up in the nest to sleep off its meal of fresh eggs.

Crows and ground squirrels can also disappear eggs. A ground squirrel may eat an egg in the nest, leaving the shell, but may just as well take the egg away, brazenly rolling it through the coop and out the pop-hole door while the chickens watch. The squirrel might then come back for seconds.

We don't have ground squirrels here in Tennessee, but we do have crows. So far they haven't ventured inside our coop to pilfer eggs, but when a crow finds a guinea nest hidden in our hayfield, it keeps visiting the nest until the eggs are all gone, at which time the guineas typically seek a new place to start a nest. Meanwhile, the crow thumbs its beak at us by leaving

▲ Foxes are among the many predators that enjoy nutritious eggs.

Minks and weasels, too, may eat an egg at the nest or carry it away. Unlike a canine, though, neither a mink nor a weasel will disturb the nest itself. Minks and weasels also are less reluctant than canines to enter a coop to pilfer eggs.

Eggshell Remains

Not all predators of the same species consume eggs in the same way, and even individual predators may use variable tactics during a single visit or subsequent visits. An animal or bird that may remove eggs without leaving any sign may just as well dine at the nest, leaving pieces of shell. Crows, ground squirrels, minks, and weasels, for example, may either remove eggs entirely or eat them on the spot. Egg size influences the decision: larger eggs for eating in, smaller eggs for takeout.

The appearance of shell pieces left in or near a nest can provide important information as to the predator's identity. Clues include whether the shell was opened on the side or at an end, the size of remaining shell pieces, whether they were left in or away from the nest, and whether or not any egg residue remains inside the shell fragments.

Hole size. Many egg thieves first make a small hole in an egg's shell. Some predators then either lick out the contents or tip the shell to drink the contents. Others make the hole bigger so they can more easily get to the contents. If more than 75 percent of the original shell remains intact, the hole is considered to be "small." If more than 50 percent but less than 75 percent of the original shell remains, the

empty shells along our driveway. Jays, magpies, and ravens are similar to crows in their appetite for eggs.

A rat may roll an egg down into its tunnel. We've not had that happen inside our coop, where we rarely see a rat, but once a rat removed eggs from a nest a duck made under an apple tree in our orchard. When we put a ceramic egg in the nest, it disappeared, too — for a couple of days, until the indignant rat shoved it back out of the tunnel entrance.

Coyotes, foxes, and domestic dogs may also pilfer eggs — usually from a nest outside the coop — without leaving any clearly visible signs. A canine either eats an egg on the spot or carries it away in its mouth. Close inspection of the nest may reveal signs of pawing if the perp was either a coyote or a dog.

hole is considered to be "large." In general, the size of the hole is relative to the size of the predator. Badgers, foxes, raccoons, and skunks typically leave shells with large holes. The wily coyote may break this rule by leaving more shells with small holes than with large holes.

Hole position. An egg normally rests on its side; therefore most egg eaters open the egg by making a hole on the top side of the shell. Birds, including chickens, typically open a shell from the side, as does a coyote or a fox. Egg thieves with dexterous front paws may tilt an egg on end and bite into the shell at the pointed end. Ground squirrels, raccoons, rats, and skunks are end openers. Many predators leave shells with a large hole that extends from the side to an end (or from the end to the side). Others leave shells with a hole both on the side and at an end, or with multiple holes. The size of the egg relative to the eater influences where the animal starts. If the predator is large in relation to the egg, it is more likely to open the shell from the side. If the egg is large in relation to the predator, the animal is likely to open the shell at the pointed end.

Shell parts. Some predators, including opossums and raccoons, may eat the contents of an egg in the nest and then eat the shell, leaving bits of shell fragments. Fragments may also be left by an animal that pecks or bites into the shell and then carries the egg away to eat elsewhere. Such fragments might not be obvious unless you deliberately look for them. Messy badgers, on the other hand, either crush eggs in their mouth and spit out the wad of shells or trample empty shells, leaving them flattened in the nest. An egg's size and shell thickness affect an animal's ability to crack the egg open, as well as the amount of shell breakage necessary to get at the contents. A ground squirrel, for example, might crush a thick-shelled egg under its body, rather than attempt to chew into the end.

Tooth punctures. A thief that pecks a shell or grasps it between its teeth may leave small punctures — with a diameter of ⅜ inch or less — on the shell that are distinct from the opening through which the egg's contents were extracted. Minks, weasels, and other small mammals open a shell by repeatedly biting into it, leaving paired punctures made by their canine teeth. Coyotes, too, may leave canine tooth marks in shells. The size of any punctures found in the shell, and their distance apart, can be helpful in determining that the predator is a mammal, how big the animal might be, and therefore what species it could be. For example, punctures left by a coyote will be at least an inch apart, while those of a mink will be about half an inch apart, and those of a weasel will be less than half an inch apart. (See Average Distance between Canine Teeth Punctures on page 261.)

Shell location. Shells left in a trail, scattered, or accumulated away from the nest — and how far they are from the nest — can provide clues. If all the shells are left in or close to a nest, the perp would typically be either a skunk or a raccoon. If a single raccoon was the raider, the shell pieces likely will be in one place; if a 'coon family had a party, the pieces will be more scattered. Both skunks and raccoons may leave a trail of shells

Egg Thief Detective Key

TYPICAL CLUES	POSSIBLE PREDATOR
Some eggs missing, crushed shells or empty shells with large holes in side, no yolk residue, nest and surrounding area trampled, evidence of digging, eggs cached nearby	**badger**
Some eggs eaten, crushed shell pieces clinging to shell membrane	**cat** (domestic or feral)
Empty, partially eaten shell or no shell, nest wet and messy	**chicken**
Some or all eggs missing, no clues or empty shell with small jagged hole or holes in side, left in nest or (more often) deposited nearby	**corvid**
Some eggs missing, no clues or bits of shell in pawed nest, some shells with small holes in the sides scattered away from nest, no yolk residue, maybe 1"-apart tooth punctures in intact parts of shells	**coyote, dog**
All eggs missing, no clues, or shells found away from nest with longwise holes in sides, no yolk residue	**fox**
Some eggs missing, no clues or partially eaten eggs, or shells with large irregular end holes with finely serrated edges scattered away from nest or trailing away	**ground squirrel**
Some or all eggs missing, no clues or shells in nest with one or more narrow longwise holes tapered at both ends, yolk residue in shell or spilled into nest	**gull**
Some eggs missing, no clues or shells with long narrow holes, fine tooth marks around hole edges and through intact parts of shells containing yolk residue, in and close to, or trailing away from, a nest near water	**mink**
Most eggs crushed, shells chewed into small pieces, yolk residue on shells and spilled into nest	**opossum**
Some eggs eaten, shells with large crushed-in round holes in end and shell fragments in or near nest, no yolk residue, shells and nest trampled	**raccoon**
No clues or hole chipped in side or end of empty shell	**rat**
Some or all eggs missing, no clues	**snake**
Some eggs eaten, shells with large crushed-in oval holes in side or end, left in or near torn-apart nest, no yolk residue	**skunk**
Eggs gradually missing over days, no clues or shells with long narrow holes with finely serrated edges and tooth marks through intact parts of shells, in and close to nest	**weasel**

Note: Likelihood of a visit by any specific egg predator depends on whether the nest is inside or outside a coop, and whether or not the coop is secured at night.

leading away from nest. Shells found well away from the nest were more likely left by a crow or a ground squirrel.

Egg size can influence shell location — smaller predators may have a difficult time moving large eggs. A crow, for example, is likely to eat a goose or turkey egg in the nest but carry away a chicken, duck, or guinea egg. On the other hand, I once saw a skunk roll a goose egg away from a setting hen's nest, while her irate mate honked in protest. The skunk's eagerness to get the egg apparently overcame its fear of the gander.

Egg residue. Some predators are adept at licking the inside of a shell clean. Messier thieves may leave some of the contents inside the shell or drip egg goo into the nest. Birds, including chickens, as well as minks may leave some of an egg's contents in the shell, while coyotes, foxes,

ground squirrels, raccoons, and skunks generally will not. When a lot of egg residue drips onto the nesting material, the work is likely that of a badger, opossum, or chicken.

Other Clues

Besides the condition and location of any remaining shells, other important clues include whether any whole eggs remain in the nest, whether eggs were removed from the nest and hidden nearby, and the extent of damage caused to the nest during the raid.

Eggs left in nest. Some greedy predators pilfer all the eggs when they raid a nest. Others take one or a few at a time and leave the rest for later. Predators that don't necessarily take all the eggs in a nest include coyotes, badgers, birds,

Typical Eggshell Clues

small hole on the side = **corvid, coyote, fox, rat**

large hole on the side = **badger, chicken, skunk**

lengthwise narrow hole on the side = **gull, fox, mink, weasel**

small hole on the end = **rat**

large hole on the end = **ground squirrel, raccoon, skunk**

crushed shell = **badger, cat, ground squirrel, opossum**

ground squirrels, raccoons, minks, and weasels. The number of eggs in the nest, the number of egg thieves at work, their size, and how hungry they were all influence whether or not any eggs will be left. A single crow may pilfer one egg at a time, but if the crow is training fledglings, the entire nest may be cleaned out in short order. Similarly, one raccoon may not polish off a nest full of eggs, but a family of raccoons likely would. A small or less hungry predator could become sated before eating all the eggs, perhaps leaving an egg partially eaten.

Some clever egg predators, including magpies and ground squirrels, attempt to hide any eggs left in the nest by covering them with nesting material. This sign is difficult to spot because it looks pretty much as if the eggs were covered by the hen that laid them. The main difference is that when a hen visits her nest, the number of eggs increases; when a predator visits, the number decreases.

Cached eggs. Some predators remove eggs from a nest and stash them for a later meal. A red fox, for instance, may cache eggs but will do so far enough from the nest that most likely you would not find the cache. A coyote, too, may stash eggs away from the nest, sometimes one here and one there.

A badger, on the other hand, will hide one or more eggs in a hole dug close to the nest and then cover the eggs with soil or debris. A sign of digging or disturbed soil and debris near a nest where eggs are missing is a pretty good indication that a badger was the culprit.

Nest wrecked. Some nest-raiding predators not only steal the eggs but also wreck the nest by pulling apart and scattering the nesting material. Animals that may disturb or destroy a nest include badgers, coyotes, skunks, and crows and their kin. A gang of raccoons may trample a nest without scattering nesting material.

Tracks and trail clues. Tracks, scat, odors, and other signs (as described in chapter 2) can be helpful in identifying an egg thief. But just because an animal was in the area does not necessarily mean that it raided the nest. On the other hand, if you find empty shells covered with fox poop or sprayed with skunky-smelling fox urine, you can be pretty certain a red fox ate the eggs and then left his calling card.

BLAMING THE WRONG CRITTER

Some predators are also scavengers that eat carrion (things that were already dead when the predator got there). Scavengers that commonly also attack live chickens include coyotes, dogs, opossums, raccoons, raptors, and rats. If you see one of these animals chewing on a dead chicken, you can't always be sure if it killed the chicken or was just taking advantage of some other predator's fresh kill.

Curiosity Killed the Goose

Geese are inquisitive creatures that check out novel things by nibbling at them. Our Embdens were fond of resting in the shade underneath our pickup truck, where they persisted in checking out the wiring. After twice replacing the truck's destroyed wiring harness, we confined the geese within a fence enclosing their pond and a small grove of pine trees.

One morning I found one of the female geese lying dead at the edge of the pond. I could see no marks on her to indicate that she had been attacked by a predator. The other geese appeared to be healthy and unconcerned, indicating that they had not been traumatized by a predator attack.

The goose's death remained a mystery until a few days later when I was mowing around the pine trees and discovered a rather large hornet's nest hanging from a low branch. The bottom half of the paper nest had been nibbled away, and bits of paper were scattered on the ground.

My conclusion is that the goose spied the strange thing hanging from the pine branch and checked it out in her usual way — by nibbling at it. Irate hornets swarmed out to pay her back for destroying their handiwork.

Hornet venom has some of the same components as bee venom, but it also contains high levels of acetylcholine, which intensifies both pain and reaction to the venom. Multiple stings most likely caused shock, respiratory failure, and death to the unfortunately inquisitive goose, and the sting marks were hidden from view beneath her thick multilayer plumage. Luckily, we haven't run across another hornet nest on our farm.

Signs of Scavenging

Crows, magpies, ravens, gulls, and even other chickens will pick at the flesh of a dead chicken. Unless you are able to inspect the body soon after the kill, the actual predator's characteristic signs may be obliterated by the scavenger, making the real perp's identification more difficult.

Crow relatives (corvids) and gulls are not strong enough to kill mature, healthy poultry, so if you see one of them pecking at a body, they are likely scavenging some other predator's kill. An exception would be if the bird being pecked had been ill or injured. Then a corvid or gull might attack the weakened bird, starting with the eyes. A sign that the bird was killed by a corvid or gull is bloody eye sockets; if the eye sockets are not bloody, the bird was already dead before the scavenger got there. Another sign that the bird was killed by a corvid is bloody debris on nearby vegetation, where the corvid wiped its beak after dining.

Hawks and eagles may also feed on carrion. By examining feathers plucked from the victim, you may be able to determine if your bird was killed by the raptor or if the raptor came along and discovered the dead bird. Feathers with a little tissue clinging to the quills were pulled from a cold, dead bird. Feathers with smooth, clean quills were plucked while the killed bird was still warm.

Other signs of scavenging include the absence of blood. Poultry fed on as carrion do not bleed. Further, scavengers do not leave the characteristic tooth or claw marks that are left by predators.

A chicken that has prolapsed may have a similar appearance to a hen that was attacked by a weasel or one of its relatives. The protruding pink tissue of a prolapse attracts other chickens to peck, and if they peck long enough and hard enough before you intervene, they will eventually pull out her intestines. A hen with bites around the cloaca and intestines pulled out may well be the victim of cannibalism within the flock.

Other signs of cannibalism that may be mistaken for signs of predation are missing toes and wounds around the top of the tail of growing chickens. A hen with slice wounds along the sides of her back got them from being repeatedly mated by a sharp-clawed rooster.

Vultures

Vultures are primarily scavengers that subsist mainly on animals that have been dead for two to three days. A vulture is a large, black raptor with an exceptionally wide wingspan. More gregarious than most other raptors, vultures often congregate in sizable groups — a fact I discovered one foggy morning when I opened my front door and was startled when a beech tree up the hill eerily exploded in a cloud of turkey vultures that had roosted there overnight.

Vultures hunt by flying in large circles, tilting this way and that while seeking something to eat. You are most likely to see them in flight during the early part of the day, up until about noon.

Vultures have weaker feet than most other raptors and therefore are less able to kill or carry live prey. On the rare occasion when a vulture does kill — which is

Turkey vulture
(*Cathartes aura*)

California condor
(*Gymnogyps californianus*)

Black vulture
(*Coragyps atratus*)

mostly when it can't find anything to eat that's already dead — it targets an animal that is seriously ill or injured and therefore easy prey.

The vulture's beak is stronger than that of most other raptors, giving it an edge for ripping into tough hides. The vulture also has a bald head and neck, which gets less messy than would a feathered head poking around in a rotting carcass.

When most people refer to a vulture, they mean the red-headed turkey vulture (*Cathartes aura*), also called a buzzard, which ranges throughout the United States and into southern Canada. Turkey vultures roost in large committees at night then break up to hunt independently during the day. They are the birds most commonly seen cleaning up roadside kills. They find dead animals by flying low and using their excellent eyesight to spot a carcass, along with their outstanding sense of smell to detect the delectable aroma of gas emitted by rotting flesh. In flight, the turkey vulture rocks back and forth as it glides gracefully on long narrow wings, which can have a span of up to 6 feet. Rarely does it flap, and when it does, the wing beats are slow and lazy. Seen

Predators That Eat Carrion

	COMMONLY	SOMETIMES
Badger		X
Bear	X	
Corvid	X	
Coyote	X	
Dog, domestic	X	
Eagle	X	
Falcon		X
Fisher	X	
Fox		X
Gull	X	
Hawk, red-tailed, and other buteos	X	
Lynx		X
Marten	X	
Opossum	X	
Owl, great horned		X
Raccoon		X
Rat	X	
River otter		X
Skunk	X	
Vulture	X	
Weasel (true weasel)		X
Wolverine	X	

from below, the wings are two-tone: black toward the front with silvery-gray flight feathers toward the back.

The California condor (*Gymnogyps californianus*) occurs only in parts of California, Arizona, and Utah. This vulture was once extinct in the wild but has since been reintroduced through the release of offspring of captured condors, though it remains quite rare. It is the largest bird in North America, with a wing span that can approach 10 feet. The condor may go for up to two weeks without eating and then gorge itself, usually on the carcass of a mammal, and only rarely on a dead bird.

The North American black vulture (*Coragyps atratus*) inhabits the southeastern United States and is smaller than the other two species. It may be distinguished from the turkey vulture by its black head, shorter tail, and short, broad wings with a span of only about 5 feet. It flaps its wings more often than the turkey vulture, following each series of snappy wing beats with a short glide — like a big old bat. The tips of the wings end in silvery feathers, like the fingers of a pair of white gloves. While the black vulture eats mainly carrion, it may also pilfer eggs or kill newborn animals, including ducklings. It does not have the same keen sense of smell as the turkey vulture, so to find a meal it must either rely on its good eyesight or else follow turkey vultures to their dining spot. Whereas the turkey vulture tends to hunt alone or in twos or threes, black vultures hunt in larger flocks that may in fact include turkey vultures.

A group of feeding vultures is appropriately called a wake. When poultry is on the menu, the wake is most likely to consist of turkey vultures eating a bird killed by some other predator. We should be happy to have vultures clean up the mess other predators leave behind.

FOUR FUN FACTS ABOUT VULTURES

▶ The three vulture species found in the United States and Canada are in the family Cathartidae, derived from the Greek word *katharsis,* meaning cleansing.

▶ Cathartids are called New World vultures to distinguish them from Old World vultures, which are in the same family (Accipitridae) as eagles and hawks.

▶ Unlike most other birds, New World vultures do not have a syrinx, or voice box, and therefore make little noise except for raspy hisses.

▶ Compared to other scavengers, vultures have stronger stomach acids, which destroy any botulism toxin or infectious bacteria that may be present in rotting meat.

2. WHO COULDA (OR COULDN'TA) DUNNIT?

"Eliminate all other factors, and the one which remains must be the truth."

Sherlock Holmes

An essential clue in attempting to identify a poultry predator is to determine whether or not the potential suspect exists in your area. Some species are extremely shy and therefore live only in remote areas. Others are more adaptable and may take up residence in urban, suburban, and rural areas alike. Some remain in the same general area year-round, while others seasonally migrate or otherwise alter their habitual routines. Some predators venture out during daylight hours, while others prowl in the dark of night. Knowing which predators are likely to be found in your area and what their routine habits are provides important clues.

If our trail cam hadn't captured this shot, we would never have known that a bear had visited our farm.

RANGE AND HABITAT

The climate you live in, your elevation, and your proximity to wooded areas or bodies of water all influence the presence or absence of different predator species. At the same time, the size of a predator's territory can vary depending on the availability of resources needed for the animal to survive and procreate. Such resources fluctuate with weather patterns (particularly temperature and rainfall extremes) and a reduction of habitat, both of which can cause an animal to expand its territorial range or move into a new area.

Therefore, a potential predator may not necessarily be one that is typically found in your area. When my husband, Allan, and I moved to our Tennessee ridgetop farm, we were told that bobcats remain close to the distant Cumberland River. So when we found what looked suspiciously like a bobcat print in the freshly tilled soil of our garden, we looked for other possible explanations. Not until we set up a trail cam did we learn that bobcats do occasionally visit our farm.

Similarly, bears are not typically seen in our area. Had we ever found sign indicating the presence of a bear, we would have sought an alternative explanation. However, again, the trail cam showed us that a bear once passed through our farm. Apparently it had other things on its mind so, thankfully, even though it wandered close to our poultry yard, it expressed no interest in our birds.

The predators that most likely come to mind when our poultry yard has been visited are bobcat, coyote, crow, eagle, fox, hawk, jay, opossum, owl, raccoon, rat, rat snake, and skunk. Ducks on our pond are additionally vulnerable to bullfrogs and snapping turtles.

Predators we are much less likely to see are bears, cougars, falcons, gulls, minks, ravens, and long-tailed weasels. Predators we don't even consider (because their range is geographically distant from us) are alligators, badgers, fishers, ground squirrels, lynx, magpies, martens, otters, ringtails, wolves, wolverines, and both short-tailed and least weasels.

To help you determine the predators most likely to live in your area, the predator profiles found later in this book include information on each species' habitat and range. Using data provided in the predator profiles, information gleaned from citizen science websites that track sightings of specific animals, and any additional information available to you about wildlife in your local area, fill in the Predators in My Area checklist on page 260 to keep track of potential poultry predators likely to visit your area.

Where Predators Are Most Likely to Be Seen

PREDATOR	URBAN	SUBURBAN	RURAL	REMOTE
House cat	X	X	X	×
Bobcat			X	X
Lynx				X
Cougar			×	X
Domestic dog	X	X	X	×
Fox, red		X	X	X
Fox, gray			X	X
Coyote			X	X
Skunk	X	X	X	X
Weasel		X	X	X
Mink			X	X
Marten				X
Fisher				X
Otter			X	X
Raccoon	×	X	X	X
Opossum	×	×	X	X
Black bear	×	×	X	X

X = typical
× = less likely

WHAT'S IN A NAME?

A source of potential confusion in determining whether or not a certain predator might be found in a given area is the fact that in different geographic locations, different names are used in referring to the same species.

Catamount, cougar, mountain lion, panther, and **puma** all refer to the big cat *Puma concolor*. Thank goodness no other animal goes by as many different names — at last count more than 80, according to Kevin Hansen in his book *Cougar*.

A **wildcat** is the same as a lynx (*Lynx canadensis*).

Brush wolf and **prairie wolf** are other names for a coyote (*Canis latrans*).

Fox, when used without a modifier, refers to a red fox (*Vulpes vulpes*) rather than a gray fox (*Urocyon cinereoargenteus*).

Timber wolf or **western wolf** is the same as a gray wolf (*Canis lupus*).

Virginia possum, or just plain **possum**, refers to the opossum (*Didelphis virginiana*).

Civet cat, miner's cat, ringtail cat, and **ring-tailed cat** all refer to the ringtail (*Bassariscus astutus*), which isn't even a cat.

A **fisher cat** is a fisher (*Martes pennanti*), which also isn't a cat, and neither does it fish.

Polecat isn't a cat, either. In North America, it refers to either a skunk (a member of the Mephitidae family) or a domestic ferret (*Mustela putorius furo*), also known as an American polecat.

American pine marten, or simply **pine marten**, refers to the American marten (*Martes americana*).

A **chicken snake** is a black rat snake (*Pantherophis obsoletus*), also called eastern rat snake, western rat snake, or simply rat snake (or ratsnake).

Ermine and **stoat** both refer to the short-tailed weasel (*Mustela erminea*).

A **chicken hawk** could be a Cooper's hawk (*Accipiter cooperii*), a sharp-shinned hawk (*Accipiter striatus*), or a red-tailed hawk (*Buteo jamaicensis*).

Pigeon hawk refers to the falcon otherwise known as a merlin (*Falco columbarius*).

Duck hawk refers to a peregrine falcon (*Falco peregrinus*).

A **hoot owl** may be either a barred owl (*Strix varia*) or a great horned owl (*Bubo virginianus*).

Buzzard in North America is a turkey vulture (*Cathartes aura*), sometimes known as a New World buzzard to distinguish it from the European buzzard, which is not a vulture but rather a hawk of the *Buteo* species.

SEASONAL PATTERNS

Predators and poultry share the same set of behavioral motivations: to get enough to eat, to avoid being eaten, and to produce offspring. As long as a predator finds the resources it needs (food, water, and safe digs), it will follow a predictable routine. Depending on the animal's appetite and the size of its territory, its routine may follow a daily cycle or may be repeated over a period of several days.

A predator's routine may change because of seasonal migrations as required to obtain the necessary resources or compensate for changing conditions such as altered weather patterns or drought (lack of drinking water). Such changes aren't incidental, because when one animal or family changes its routine, others may move in to take advantage of reduced competition.

Most predators raise their young during the spring and summer months, typically choosing an area that offers plentiful resources to tap into while feeding young and, later, while teaching them to hunt. If those resources include your poultry, the same animal(s) may come back repeatedly for an easy meal. At other times of year, visits to a poultry yard may be less predictable, since many predators roam over a wider territory after their young are old enough to leave the nest or den.

With the exception of owls, northern raptors migrate southward during the fall and winter, during which they may visit a given area for a week or two before continuing their journey. The same species living in a more hospitable southern habitat may remain in their home territory year-round. As a result, the native southern population can increase dramatically in fall and winter, thanks to an influx of migrants from the north.

Among corvids, some migrate and some do not, with a similar resulting ebb and flow of populations. Magpies and other species typically found at high elevations are less likely to migrate southward, but during the cold months they may move to a lower elevation in the same region.

Mammals don't migrate, in the sense of traveling long distances as the seasons change, but some species seasonally expand their territories. Predicting when a predator might visit your poultry yard therefore requires knowing not only its typical breeding pattern, but also whether the predator lives in your area seasonally or year-round.

Commonly known among poultry keepers is that domestic dogs, along with certain other predators, kill for the sheer mirth of it — called sport killing, joy killing, or thrill killing. What else would you think if you went out to visit your poultry only to find multiple birds dead but not carried away or even eaten? Such wanton killing is enough to send you on your own predator killing spree.

But wait. It isn't a simple case of an animal run amok. Let's try to figure out what's going on in the predator's head.

Biologists call it the henhouse syndrome, as in "the fox got into the henhouse." More technically it's termed surplus killing. When predators encounter easily accessible domestic prey, they initiate surplus killing — they kill animals without eating them. L. Scott Mills describes the phenomenon in his book *Conservation of Wildlife Populations* as an almost inevitable result of a high-performance predator confronted with an easy target.

As Mills explains, predation involves four distinct behaviors — search, pursue, kill, consume — that are independently reinforced. That means the predator doesn't need to complete the entire predation act from search to consume to feel satisfaction but receives some sort of reward for successfully carrying out each individual behavioral step. Think about a

young carnivore that is not yet adept at hunting but learns to hunt by receiving psychological encouragement for playing with dead or live prey provided by its parents.

Each step from search to consume presents challenges and therefore can be time- and energy-consuming to complete, and success is by no means guaranteed. As Mills points out, when a predator is presented with a situation where the first two steps, search and pursuit, are made ridiculously easy — such as finding poultry confined to a coop or yard — the third step, killing, can be done with little risk of injury. The predator therefore has no adaptive reason to stop killing, regardless of whether or not it eats the prey.

Not only that, but as one of the four distinct steps to predation, killing must provide its own reward. In other words, it gives the predator a rush (yes, killing for thrills, especially when birds panic, adding the element of brief pursuit to increase the fun). This drive to kill for sport contributes to the animal's survival, for with each kill the predator gains experience that may prove invaluable in the future.

Further, surplus killing allows a predator to provide food for its young and is also a way to save food for times when resources might be scarce. When a predator leaves dead chickens or other poultry scattered around the coop or yard, it has no way of knowing that when it comes back later, the dead birds will have been removed. In such a case, the predator has basically wasted its time and must search for other prey to kill. If it didn't kill the entire flock on the first visit, it may well come back and kill the rest.

Poultry predators that engage in surplus killing include members of the dog family, cat family, and weasel family. All of these animals are known to cache food for future use, just as we humans tend to buy more than we need for tonight's dinner and cache the surplus in the fridge or freezer for a later meal. Most of us who raise poultry for meat find it much more efficient to butcher several birds at a time and freeze the surplus, rather than pluck a single bird for each meal.

Surplus killing is most likely to occur when bad weather is on the way and prey becomes scarce — typically as winter approaches. Killing an entire flock of poultry provides more meat while using less energy (in other words, it is more efficient) than hunting down individual prey. So when a predator finds a flock of poultry confined to a coop and run, can we blame it for seeing the birds as a bonanza buffet with takeout on layaway?

DAILY PATTERNS

Each animal species follows habitual daily patterns of activity and sleep. These patterns are usually regulated by sunlight or lack thereof. An animal's normal patterns are not set in stone but may change seasonally or because of internal or external influences.

Day and Night

Diurnal species tend to be active during daylight hours and sleep during the night. The word diurnal comes from the Latin word *diurnalis,* meaning daily. Diurnal species find food by relying on detailed color vision in bright light. We humans are basically diurnal. The predators we are most likely to see in action, therefore, are other diurnal species, such as corvids, gulls, ground squirrels, and raptors (excluding owls). Like humans, other animals with eyes designed for daylight vision don't see well in dim light or in the dark.

As daylight fades, so also does the color perception of diurnal predators, some of which turn in at sundown, while others continue to hunt into dusk. Meanwhile, other predators are just waking up. Vespertine animals tend to be most active during the twilight hours between sundown and nightfall. The word vespertine comes from the Latin *vespertinus,* meaning of the evening. The word *vespertine* might be easily remembered if you think of the related word *vespers* (evening prayer). Vespertine predators take advantage of vespertine prey while avoiding competition with animals that aren't active until after dark. The black bear is basically vespertine, unless disrupted by human activity; for safety's sake, bears become nocturnal in urban areas, and elsewhere during the fall hunting season.

Nocturnal animals sleep during daylight and prowl after nightfall. The word nocturnal derives from the Latin word *nocturnalus,* meaning belonging to the night. Some nocturnal species become active during dusk and continue through the night. The eyes of nocturnal animals are optimized for seeing in low light, and some also have highly developed senses of hearing and smell. The owl, for example, sees well in the dark but not so well in bright light, and it also has ears asymmetrically positioned to pinpoint the precise source of a sound. Other nocturnal animals include foxes and coyotes. Raccoons are considered to be truly nocturnal, in that they're rarely seen out and about during the day, although when one of our local raccoons lost a foot in a neighbor's trap we often saw it foraging in the daytime. Coyotes also are primarily nocturnal, based on the many images captured on our trail cam, but pups will occasionally hunt during the day. One fall we saw a young coyote in broad daylight, getting tipsy on wild persimmons fermenting on the ground beneath the tree.

In the early-morning hours between first light and sunup, matutinal predators are waking up. The word matutinal derives from the Latin *matutinus*, meaning pertaining to morning. The word *matutinal* may be easily remembered if you think of the related word *matins* (morning prayer). Matutinal predators become active at dawn, after nocturnal animals have called it a night, but before diurnal critters become active. They thus take advantage of early-rising prey while avoiding competition with diurnal predators. Few, if any, poultry predators are strictly matutinal.

Twilight Zones

Some animals hunt primarily during the twilight hours of both dawn and dusk, rather than during full daylight or the dark of night, making them both vespertine and matutinal. Such animals are called crepuscular, from the Latin word *creper,* meaning dusky or gloomy. A crepuscular animal is active just before the sun rises or just after it sets — time periods when the Earth's lower atmosphere reflects rays of the rising or setting sun and diffuses just enough light for creatures to see, but not see well. Like nocturnal species, crepuscular species rely not just on sight but also on sound, as well as smell, to find food.

Examples of crepuscular animals include cats, dogs, opossums, rats, river otters, snakes, and skunks. Crepuscular predators are often mistakenly considered to be nocturnal, since their peak activities occur at about the same time that nocturnal species are gearing up and then again when nocturnal animals are calling it a night. Also known as bimodal (from the Latin words *bi,* meaning two, and *modus,* meaning mode), truly crepuscular animals do not hunt after dark. Nor do they hunt in daylight. The theory is that crepuscular predators have just enough light to see their crepuscular prey, while feeling safe from being preyed upon by nocturnal or diurnal species. In desert areas, crepuscular animals avoid both the blistering midday heat and the chill of night.

If you have a house cat, you know that cats are crepuscular. Our cat sleeps on the couch all day and on our bed all night but prowls in the evening and annoyingly begs to play in the wee hours of morning. Similarly, bobcats and cougars are crepuscular, being most active for about three to four hours in the morning and then again in the evening. Most of the bobcat pictures captured by our trail cam are shot in the early morning or late evening, although a few occur in the middle of the night. And we once watched a young bobcat cross our backyard in full daylight.

Rule Breakers

Because they so frequently break the rules, members of the cat family (including cougars, bobcats, and lynx), as well as coyotes and black bears, might more properly be classified as cathemeral, a word that derives from the Greek *kata,* meaning through, and *hemera,* meaning day. In other words, cathemeral animals are active throughout the day (24/7). Another term for this activity pattern is *arrhythmic,* since their irregular activity may find these species out at any time of day or night. Truly cathemeral predators include bullfrogs, weasels, and wolverines.

Crepuscular or even diurnal predators that pretty much stick to their normal activity periods may be enticed to roam on a brightly moonlit night. Similarly, crepuscular or nocturnal predators may prowl on a darkly cloudy day. Most predators, as well as other animals, are flexible enough to adjust their activity times if need be. Consider the human as a prime example. Although we are basically diurnal, technologies such as artificial lights and warm clothing allow some of us to become decidedly nocturnal while others tend to be cathemeral.

Many conditions can cause an animal to be active outside what is considered to be the normal behavior pattern for its species. Nocturnal mothers typically forage during the day while feeding young to avoid leaving their defenseless babies where other nocturnal predators might prey on them. Other things that can upset the applecart include human activity (such as hunting, or the lights and noise that come with urbanization), decreased availability of preferred prey species (because of such things as loss of habitat or drought), or increased competition with the same or other species. I once saw an army of rats at midday, crawling all over a dumpster behind a local doughnut shop, enjoying the plentiful availability of scraps while avoiding competition from fellow rats that remained nocturnal. (And, no, I didn't stop for a doughnut!)

Finally, don't discount the possibility that a predator failing to observe the typical activity pattern for its species may have rabies or some other debilitating illness. Foxes, raccoons, and skunks in particular are considered to be primary carriers of the rabies virus.

A sick animal may look scruffy. It may have missing or matted fur, or weepy or crusted eyes. It may appear disoriented. It may act unusually bold or aggressive. It may tremble, stumble, stagger, walk in circles, or simply be unconcerned about remaining completely exposed.

One sunny morning I stepped out the back door of our house to encounter a pretty little red fox sitting quietly on the lawn. I made a loud noise and waved my arms to shoo it away, but it just sat there staring at me. You can bet I beat a hasty retreat and called for help.

Poultry Patterns

Chickens, turkeys, and guinea fowl are matutinal and diurnal. They generally go to roost at sundown and remain there until first light. Unless they roost inside a secure coop, during the night they are vulnerable to vespertine and nocturnal predators. A flock that sleeps protected indoors overnight is vulnerable only to crepuscular and diurnal predators.

Ducks and geese tend to be cathemeral and therefore do not take kindly to being cooped up at night, making them susceptible to all manner of predators. A pond, no matter how large, does not provide sufficient safety. It won't protect waterfowl from a raptor or from a land-based predator that is a skillful swimmer.

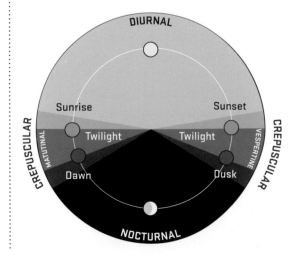

Predators and Their Period(s) of Greatest Activity

	DAY (diurnal)	DAWN & DUSK (crepuscular)	DAWN (matutinal)	TWILIGHT (vespertine)	NIGHT (nocturnal)	CATHEMERAL (arrhythmic)
Alligator		typical				
Bear, black	summer			yes	fall, and urban	possible
Bobcat	rare	typical				possible
Bullfrog						yes
Corvids	yes		yes			
Cougar		typical				possible
Coyote					yes	possible
Dog, domestic		typical				
Fisher		typical			typical	
Fox					yes	possible
Ground squirrel	yes					
Gulls	yes		yes			
Cat, domestic or feral		typical				possible
Lynx		typical				possible
Mink	possible	typical			typical	
Opossum		typical				
Owl	possible	typical			possible	
Raccoon					typical	
Raptors (except owls)	yes		yes			
Rat	possible				yes	
River otter		typical				
Weasel, short-tailed (ermine)				no	yes	possible
Skunk		typical				
Snake		typical				
Snapping turtle					yes	
Weasel	summer				winter	yes
Wolverine					yes	

PREDATORY CYCLES

Wildlife populations tend to increase and decrease on a regular basis. When the natural prey of a particular predator are plentiful, the predator is likely to go after the easy prey and leave your protected birds alone. However, predator populations tend to increase during times of plenty, and as the predator population increases they make inroads into the prey population, gradually reducing their available food supply.

Predators that regularly wander past your poultry yard may suddenly, out of hunger, stop and eye your birds. Seemingly out of nowhere, your flock is attacked. After a prolonged period without any sign of predation, it's easy to see why poultry keepers may believe their birds are safe from local predators, only to be puzzled by a sudden attack. Unfortunately, once a predator knows what's on the other side of the fence, it is likely to come back and, in some cases, train its offspring to do likewise.

How often a predator needs to eat, or provide food for young, is another determining factor. A daddy fox feeding his mate and pups brings food to the den every four to six hours. A raccoon, on the other hand, typically returns for a meal of fresh meat about every five days.

Local predators that go hungry or are unable to feed their young will eventually either die of starvation or move to better hunting grounds, reducing the possibility of an attack on your poultry. But as the predator population decreases, the population of their prey increases, starting the cycle all over again. Some predator/prey cycles occur slowly over many years. Others occur more rapidly.

The cycles can be affected by many different outside influences. One such influence might be a disease spreading through the population of either the predator or the prey. Another might be multiple predator species seeking the same prey. Or populations of either predator or prey may be influenced by weather conditions, especially temperature extremes or drought.

Another type of cycle is the annual cycle of producing young. As the young grow, the parent must provide more and more food until the kids are able to hunt on their own, at which time the young ones may or may not move along to a different territory. By studying the biology of your local predators, including their typical number of offspring and their preferred prey, you can learn to predict approximately when you might expect certain predators to attack your poultry yard.

> For more about the lives of many predators discussed in this book, follow this link to a supplementary resource on the Storey Publishing website: https://www.storey.com/poultry-predators/

The 80-Hen Heist

Within a Massachusetts campground lived 81 hens as part of a petting zoo maintained by my friend Paul. The constant presence of family campers served to protect the chickens from predators. Suddenly, and mysteriously, two hens vanished. The next day, two more disappeared. And so it went, day after day.

Paul was baffled, until one day he spotted a huge fisher in broad daylight climbing over the poultry fence carrying one of his hens. He attempted to pursue the fisher, whereupon she dropped the hen and turned to face him. "It was a scary moment," recalls Paul, who cautiously backed away. To his relief, the fisher picked up her prize and continued on her way.

The fisher persisted in pilfering two hens a day until Paul's once sizable flock was down to a single hen. At that point either the fisher decided one hen wasn't worth the trouble, or her kits had learned to feed themselves, because just as suddenly as she had started pillaging poultry, she stopped.

3. | SLEUTHING FOR CLUES

> **"In an investigation, the little things are infinitely the most important."**
>
> Sherlock Holmes

Each predator leaves a unique calling card that, when carefully read, can tell you what kind of animal was involved. Some signs are conspicuous; others may be less obvious. Having raised chickens for some 50 years, first in suburbia and now in a relatively remote area, I have seen quite a few of these signs. Even so, every now and then I get stumped, mainly because not all predators have read the manual, so they don't always conform to their own species' standard operating procedure. Still, anyone who makes an effort to be observant can spot certain tell-tale clues that point in the general direction and, considered together, can lead to a confirmed identification.

EYEWITNESS EVIDENCE

Seeing a predator in action is, of course, the quickest and surest way to identify it. I never would have known where my little black bantam hen disappeared to had I not been visiting the flock when a red-tailed hawk swooped down into the yard and swiftly carried her away.

But seeing is not always believing. Late one night I was awakened by a commotion in my yard. When I ran to the window, I saw what looked like a short man reaching into my elevated outdoor pen. When I snapped on the yard light, the person appeared to double over and run down the driveway into the dark. By the time I was fully awake, I realized that I had seen a rather large German shepherd standing on its hind legs while breaking into the pen.

On another occasion a great horned owl was regularly picking off young guineas that persisted in roosting on a utility wire (see Things That Go "Thump" in the Night on page 158). One night I heard a noise on the roof above our bedroom. I looked out the window and by the light of the moon saw what looked like a pudgy kid with baggy pants standing in the yard. Both events indicate that nighttime light and shadows can easily distort perception.

The next best thing to witnessing predation in person is to catch the predator on a motion-activated camera. If the area you want to monitor is close to electricity and an internet connection, you might opt for an outdoor home security camera that sends images of detected prowlers to your smartphone or email address. Important features are motion detection and good night vision. Some models are also Wi-Fi capable. Reviews of the latest brands and models may be readily found online.

Where power and internet are not handily nearby, you might opt for a motion-detecting game camera, also known as a trail camera or trail cam. Lots of models are available with a wide range of prices and features, including time lapse and video options. These cameras store images on a memory card, and some also use cell technology to send images to your smartphone or email address. As with security cameras, reviews of current models may be found online.

Because our flock is not within convenient range for a security camera (even with Wi-Fi boosters), we monitor our poultry yard with a game camera. My favorite website offering comparative game camera information is Trailcampro (www.trailcampro.com), which does its own annual comparative tests, lists the good and not-so-good features of each brand, offers unbiased comparisons, and includes lots of sample photos from each model, mostly submitted by customers. The site also has a trail camera selection guide that takes you step-by-step through the process of deciding which camera would work best for your situation. If you still have questions, you can call and speak with someone who has firsthand knowledge of the camera that interests you.

▲ The trail cam captured this bobcat returning for another turkey.

▲ Royal Palm turkey Tom White struts his stuff in front of a trail cam.

We purchased our first trail cam when our Royal Palm turkeys started disappearing (see The Case of the Vanishing Turkeys on page 194). Pretty quickly we learned something about operating a trail cam that isn't in the user manual: If you don't want eight gazillion pictures of your own poultry, turn the camera off after sunrise and back on before nightfall.

The first day our new cam was set up, our Royal Palm turkey Tom White spent the day strutting back and forth in front of it — to the tune of nearly 3,000 frames. With the trail cam on only at night, we saw that our poultry yard was regularly visited by an opossum, a skunk, a rabbit, and an early-rising crow. Then, finally, the cam captured our perp — a bobcat — coming back for another turkey.

MAMMAL SIGNS

More often than not, poultry predators prefer to work when no one is around to catch them in action. And they sometimes manage to evade a trail cam's limited field of view. In such cases, identifying a predator requires assessing the signs it leaves behind. The most obvious signs to look for are tracks and scat.

Those clues, however, don't necessarily identify an animal as the predator in question. Maybe it just happened to be in the area where predation occurred. Tracks and scat must be considered along with other signs, including those described in chapter 1.

Mammal Tracks

Although each species produces tracks with distinctive features, learning to identify a predator by its tracks takes some degree of knowledge and experience. Tracks among animals of the same species can be as varied as the sizes and shapes of human hands or feet. Tracks made by walking are not the same as tracks made by running. And the surface on which an animal leaves tracks also influences their appearance. Yet, given good, clear tracks, an experienced tracker can identify an animal's species and possibly more — such as its age, gender, and state of health.

Tracks are clearest when made in thick mud, moist clay, fresh snow, wet sand, or soft, sandy soil. They're also pretty clear in freshly poured concrete, as we learned when we added a concrete pad outside our pullet house. That evening when we checked the hardening concrete, we found it covered with raccoon prints.

On hard or frozen soil, you might spread flour or cornstarch on the ground where the predator will likely step, such as directly in front of the pophole. This method requires persistence if you're dealing with a predator that comes around irregularly. If the flour is likely to be disturbed during the day by poultry activity, you'll need to smooth it or add more each evening after the birds have gone in for the night.

Finding clear tracks that haven't been obliterated by active poultry can present a definite challenge, especially if the flock is released in the morning by an automatic door before you get there to examine the ground. In such a case, you might want to adjust the door opener to keep the birds confined for longer than usual in the mornings, giving you time to search for tracks.

Taking photos can help you identify tracks, as well as give you a record of when and where you find tracks. Among my camera gear I keep a quarter coin, because it is nearly 1 inch in diameter. When shooting tracks, I always include the coin in at least one photo to show the track's scale. The problem with photographs is that shadows can obscure subtle details, and getting adequate depth of field can be difficult.

If you draw better than I do (which nearly everybody does), instead of a camera you might carry a notebook and small ruler. Sketching tracks in the notebook would allow you to include details on the page, such as when and where the track was found, along with exact measurements (such as the length and width of the track, as well as the stride and straddle of the trail, as described on page 54).

Some people enjoy preserving the best tracks they run across by casting them in gypsum plaster, and some tracking experts will help attempt to identify an animal based on a good track cast. Numerous websites describing how to make such casts may be found online by searching the key words "animal track casting."

A field guide is nice to have along when you are trying to identify an animal track. Several handy smartphone apps are available.

Track Size and Shape

A track's features, including size and shape, can give you a good idea of the animal's family group. Mammal tracks can often be identified by the following features:

- Shape of paw print

- Size of paw print

- Number of toes

- Visibility of claw marks

- Size of paw pad

- Shape of paw pad

- Trail pattern

CAT, DOG, FOX, OR WEASEL?

When you observe a track, you will notice that the toes and paw pad sink into the ground leaving raised areas between the toes and between the toes and the paw pad. If you make a cast from the track, the toes and pad will be raised and the space around and between them will be indented.

The toes (and claw marks, when present) and paw pad are considered to be the positive parts of the track. Just as important in identifying a track is the shape of the area surrounding the positive parts and giving them definition. The area between the positive parts is the negative space.

One of the ways to determine whether the track was made by a cat, dog, fox, or weasel is to look at the shape of the negative space. If the print was made by a member of the cat or weasel family, the negative space will be curved into an arch or sideways C shape (the weasel track will likely show claw marks, while the cat track typically will not). If the print was made by a coyote or red fox, the space will be in the shape of an X — roughly creating a small pyramid in the middle of the track. If the print came from a domestic dog or a gray fox, the space will be in the shape of an H. These shapes are particularly pronounced if the track was made by a front foot.

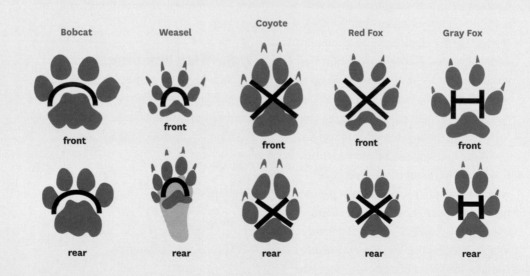

The track's shape is determined, in part, by the animal's stance: whether it travels flat-footed or on its toes. Those that walk flat-footed are called sole walkers (technically plantigrades). Sole walkers have five toes. In addition to us humans, sole walkers include badgers, bears, opossums, otters, raccoons, rodents, skunks, and wolverines. Animals that walk on their toes are called toe walkers or ball walkers (technically digitigrades). Most toe walkers have four toes, with the exception of true weasels (which have five toes). Toe walkers include members of the cat and dog families and birds that walk on land.

A track's size can give a general idea of the animal's size, helping you distinguish between tracks of similar shape, such as a bobcat and a cougar, or a fox and a coyote. On the other hand, tracks made in mud tend to be larger than the same tracks on dry soil. For accurate assessment, measure as many tracks as you can find. Further, a large fox (male) and a small coyote (female) produce nearly identical tracks, as do a lynx and a cougar, leaving you to search for other clues.

Characteristics of Sole Walkers and Toe Walkers

	SOLE WALKER (PLANTIGRADE)	TOE WALKER (DIGITIGRADE)
Stance	Flat footed	Ball of foot
Number of toes	5	4
Stride	Shorter	Longer
Speed	Slower	Faster
Paw sizes	Rear larger than front	Rear similar to front
Legs	Shorter	Longer
Stealth	Less stealthy	More stealthy (quieter)
Stability	Good balance standing on rear feet	Poor balance standing on rear feet
Weight-carrying capacity	Greater	Less
Power to jump	Long distances	Shorter distances
Predator	Badger, bear, opossum, raccoon, rodent, skunk, wolverine	Bird, cat, dog, true weasel

Mammal Track Key

FOUR TOES

CAT FAMILY
- Round
- No claw marks
- M-shaped paw pad
- Asymmetrical

DOG FAMILY
- Oval
- Claw marks
- Triangular paw pad
- Symmetrical

WEASEL FAMILY
- Long claws
- Inside toe may not show
- Asymmetrical

FOUR TOES IN FRONT, FIVE TOES IN BACK

RODENTS
- Small prints

FIVE TOES

SKUNKS
- Front: small and round with long claws
- Rear: like a human foot

RACCOONS
- Like a human hand with long parallel fingers

OPOSSUMS
- Toes splayed
- Rear: opposable thumb

Trail Pattern

The pattern in which tracks are laid down provides an additional clue to identifying an animal, or at least its family group. Trail patterns are determined by the animal's way of traveling, or gait. Predatory mammals typically fall into one of the following four groups according to their most common gaits.

Walkers and trotters. These animals are sometimes called diagonal walkers. They produce a single line of diagonal, or zigzag, prints. This trail pattern is made by members of both the cat family (house cat, bobcat, cougar, lynx) and the dog family (coyote, domestic dog, fox, wolf) — all are four-toed animals, with a body length from shoulder to rump that is approximately the same as the length of their legs.

Cats walk more than trot, moving first the front foot on one side of the body, then the rear foot on opposite side, then the other front foot, then the opposite rear foot. The cat places its rear foot directly on top of the print made by the front foot on the same side, an orientation called direct register. The tracks appear as only two marks, arranged in a relatively straight line known as a single walk or perfect walk.

Coyotes trot more than walk, moving the front and opposite rear feet together. A trotting coyote also produces direct register tracks. When a coyote walks, however, the rear foot falls just behind, or overlaps, the front foot track, an orientation called indirect register. Indirect register tracks appear as four distinct marks.

Fox tracks are direct register, creating a trail pattern that appears more catlike than doglike.

Waddlers. An animal that waddles creates two parallel lines of tracks by moving both feet on one side of the body before moving both feet on the other side. The resulting tracks may be alternating (the rear foot steps nearly in the track of the front foot) or the front and rear foot tracks may appear side by side. In most cases the rear feet are toed inward. This pattern is made by slow-moving animals with heavy bodies, short legs, and front feet smaller than rear feet — badgers, bears, opossums, raccoons, and skunks.

Bounders. These prints appear in irregularly spaced pairs or bunches, a pattern created by a series of bounds made by crouching, pushing off with both rear feet, landing on the front feet (one slightly ahead of the other), then lifting the front feet and landing the rear feet on top of the front foot tracks. This trail pattern is characteristic of members of the weasel family (including fisher, marten, mink, and otter) — all small animals with long, narrow bodies and short legs.

Hoppers. Animals that hop have back legs that are longer than the front legs, creating tracks in usually evenly spaced bunches. Sometimes called gallopers because of their three-beat gait, hoppers push off with the rear feet and land on the front feet, one slightly before the other, then swing the back legs around and out to the front, leaving tracks of the larger rear feet ahead of the smaller front feet. Most hoppers — including groundhogs, rabbits, and tree squirrels — typically don't bother poultry or their eggs, although their tracks might be found in or near the poultry yard. The hoppers to watch out for are ground squirrels and rats. You can distinguish a ground squirrel's tracks from those of a tree squirrel by noting whether the front feet land diagonally or side-by-side; if diagonally, it's a ground squirrel.

Although these four traveling gaits result in trail patterns typically made by each species, no animal consistently travels using the same gait. When a walking cat or a trotting fox, for example, spots prey and pounces, it becomes a bounder, as does any cat or dog while running. Rats typically travel by hopping but walk when stalking, bound when pouncing, and waddle when bored.

GENDER AND AGE OF A WALKER OR TROTTER

In the dog family, the male brings food for the female and pups and helps protect them from predators. In the cat family, the male plays no role in caring for the young, and therefore the female is left to feed and care for the offspring. Although she would normally be a crepuscular hunter, the mother cat may hunt during daylight hours to avoid leaving her offspring vulnerable to predation. Where poultry pickings are easy, a mature male dog (coyote or fox) or a mature female cat is likely to return repeatedly.

When examining the trail pattern of a member of the dog or cat family, you can determine whether the animal is a male or female and mature or immature. Mature females tend to be wider at the pelvis for greater ease in giving birth, while males tend to be wider at the shoulders. For most members of the dog family, the female's rear tracks appear outside the front tracks, while the male's rear tracks appear inside the front tracks. If the animal is not fully mature, the rear track will appear neither outside or inside but directly behind the front track.

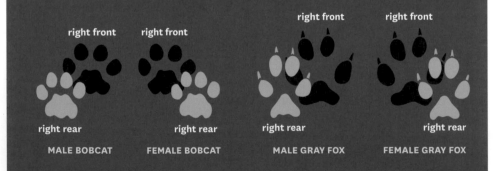

right front	right front	right front	right front
right rear	right rear	right rear	right rear
MALE BOBCAT	FEMALE BOBCAT	MALE GRAY FOX	FEMALE GRAY FOX

For foxes and cats, the smaller rear track overlays the larger front track to the outside for females and to the inside for males. If the animal is not full grown, the rear track will appear directly on top of the front track.

How Animals Move

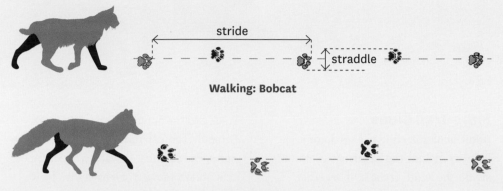

Walking: Bobcat

Trotting: Red Fox

Walkers and trotters are four-toed animals with legs about the same length as their bodies — members of the cat or dog family.

Waddling: Raccoon

Waddlers have fat bodies and short legs, with the front feet smaller than the back feet — bears, opossums, raccoons, and skunks.

Bounding: Weasel

Bounders have narrow bodies with short legs — fishers, martens, minks, otters, and weasels.

Hopping: Rat

Hoppers have rear legs much longer than the front legs — rodents, including ground squirrels and rats.

More Trail Clues

Additional trail clues to a predator's identity include the length of the animal's stride, the width of the trail, whether the prints move purposefully in a single direction or wander, possible marks left by dragging body parts, and other signs left along the trail.

Stride. The distance between tracks is not measured exactly the same way by all wildlife trackers. Some measure step as the distance between two consecutive tracks on the same side. This distance is sometimes also called stride. Another definition of stride — one that makes more sense in terms of consistent measurement — is the distance between one track and the same point on the next track made by the same foot. Stride can tell you how large the animal is (or at least how long its legs are) and how fast it is moving. A walker that breaks into a trot, for instance, has a longer stride. Stride always should be measured at the same spot on the track, whether at the heel, middle, or front of the toes.

Straddle. The width of an animal's trail pattern, or straddle, is also not defined exactly the same by everyone. The definition that gives you the most amount of information is that straddle is the distance between the outside of the left track and the outside of the right track. Straddle, too, can indicate an animal's size and speed of travel. A walker, for instance, that breaks into a trot leaves a trail with a narrower straddle, because the animal can move faster with its feet closer together under the center of its body.

Meandering. A trail that meanders, as if the animal is searching for something on the ground, provides a bit of a clue as to the animal's identity. Skunks meander, while the trail of a weasel with similar-looking tracks is more likely to be in a straight line. Bobcats, lynx, and domestic dogs also meander, in contrast to the typically straight trails of coyotes, foxes, and wolves, which travel in a straight line as if knowing precisely where they are headed.

Sliding. A few animals alternate traveling on their feet with sliding on their bellies. Sometimes they paddle while sliding, leaving paw prints along both sides of a smooth slide. Minks and river otters, both semiaquatic animals, are sliders. Minks slide in snow on downhill slopes. Otters slide down muddy or snow-covered hills but may also slide on snowy level ground, up gentle slopes, and on ice. The otter, being much larger than the mink, leaves a deeper, wider slide.

Tail dragging. Some animals hang their tails down while they travel, leaving lines in the middle of the trail pattern or, in some cases, covering tracks. Tail draggers include ground squirrels, opossums, otters, and rats.

Foot dragging. Tracks that show signs of foot dragging were likely left by a red fox or a wolverine. Foot drag marks are particularly distinct in snow.

Hair. Fur or hair caught in a fence or elsewhere can sometimes give you a clue as to a predator's identity. I once identified a neighborhood dog that broke into our barn based on clumps of hair left on a gate. If you look hard enough and have the patience for it, you might find single hairs lying on the ground.

Scrapes. Claw marks, known as scrapes, associated with scat indicate a member of either the cat or the dog family. Unlike house cats, which usually bury their scat, other felines don't always bury their scat but more often scrape together a mound of soil with their rear feet and deposit scat on top. A dominant canine will also scratch soil, using both front and back feet, near where it deposits scat. In both cases the scrape's size offers a clue as to the animal's size.

Claw rakes. Another kind of scrape is a claw rake. Just as house cats stand on their hind legs to scratch furniture, cats in the wild scratch on tree trunks or stumps to sharpen their claws and shed old claw sheaths. The height of the scrape indicates the size of the cat. A cougar, for instance, may scratch as high as 8 feet. Bears, too, leave claw rakes, although their scrapes are usually larger and damage more bark than a cat's.

Scent marks. Wherever they travel, members of the cat and dog families leave an odor from scent glands on the bottoms of their feet, conveying information regarding their age, gender, state of health, and reproductive status. Fishers and martens have similar scent glands on their hind feet. We humans may not be able to detect all the odors left by animals, but we can certainly detect some of them. Canines and felines have scent glands on both sides of the anus that can release some pretty strong odors in their poop and pee. Lynx, for instance, release an odor that smells like house cat pee, but much more pungent. Gray fox urine smells musky. Red fox urine smells skunky. Skunks, of course, are notorious for their odor, but they don't routinely spray; they reserve their odiferous means of defense for times when they feel truly threatened.

Because scent marks degrade over time, a predator may periodically return to refresh its mark. Gray wolves, for instance, refresh their territorial scent marks about every three weeks, particularly where competing wolves are likely to intrude. Females cats tend to scent-mark more often when coming into heat and male cats typically scent-mark more often when in the vicinity of a female in heat or about to come into heat. The science of analyzing scent marks (called ethochemistry) to understand their exact composition and the information they convey is still in infancy.

Scat. Many predators leave scat along their trails as a way to mark their territory and discourage competitors from sharing the available food, water, shelter, and mate(s). Trail-marking scat is more often left by males than by females.

Mammal Scat

Scat — feces, poop, poo, droppings, excrement, or whatever you choose to call it — can provide a significant clue to an animal's identification. Depending on the predator, scat may be less easy to find than tracks, as well as less easy to identify, especially since scat from a single animal isn't always the same year-round but varies in shape and consistency with the animal's diet.

Still, scat's form and dietary indicators, along with where it is found, can be helpful in attempting to identify an animal's species. For example, we often find scat in the middle of our lane, at its juncture with the driveway leading to behind our barn. Bobcats, coyotes, and foxes typically mark trails in this manner. All three species produce segmented scat, but bobcat and coyote scat is up to 4 inches long, while fox scat is up to 2 inches long. So we identify the scat as either bobcat or coyote. Bobcat scat is generally segmented and blunt at both ends, while coyote scat usually is not segmented, is tapered at one end, and contains more visible hair. However, scats are not all alike, and sometimes coyote scat is segmented or bobcat scat contains more hair than usual. Coyotes prowl our property more regularly than bobcats do, so when it's a toss-up we assume it's coyote. A conclusive decision often comes by reviewing recent images captured by our trail cam.

As with identifying tracks, taking photos of scat (including at least one with a quarter or other scale indicator) can help you identify it, as well as providing a record of when and where you found it. The mammal predator profiles in part 2 include illustrations of typical scat, along with indications of size.

Scat Shape and Size

The scat's shape is a clue as to the animal's family group because the shape of scat is affected by the diet of each species. Round pellets, for instance, are mostly deposited by herbivores — such as a deer or rabbits — which are not poultry predators. Small, oval or tubular scat that is less than 1 inch in length likely comes from a rodent, the most significant species to threaten poultry being ground squirrels (out West) and rats (everywhere).

Cylindrical or tubular scat that is greater than 1 inch in length is more indicative of a meat eater, such as a bobcat, coyote, fox, or raccoon. Carnivores can be further identified by features such as these:

- Whether the scat is flattened or rounded

- Whether the scat is smoothly cylindrical or twisted

- Whether the ends are blunt or pointed

- Whether or not the diameter varies in thickness

- Whether or not the scat is segmented

In terms of scat, the word segmented refers to a cylinder broken at right angles, creating a series of shorter cylindrical pieces, something like marshmallows. Segmented scat is usually deposited by members of the cat family — bobcats, cougars, and lynx. Bears, too, may produce segmented scat, but it is significantly greater in size than cat scat and therefore not likely to be misidentified. Segmented scat tends to be blunt at both ends.

Raccoons also deposit scat that is blunt at both ends, but it consists of a series of unsegmented cylinders of uniform thickness. Skunk scat, on the rare occasion it is found, tends to have blunt ends but consists of small irregular-shaped crumbly pieces. Also rarely seen, opossum scat tends to be pointed at one end and consists of irregular smooth pieces, not crumbly ones like skunk scat.

Scat deposited by coyotes and wolves differs from cat scat in several ways. It is rarely segmented, and it is usually blunt or rounded at one end and pointed or tapered at the other end. The pointed end may trail off into a long, thin curl. Fox scat is pointed at both ends and consists of a ropy series of connected pieces.

Most predators produce scat consisting basically of round cylinders. Flat scat, on the other hand, is deposited by members of the weasel family. Their scat has the appearance of flattened, twisted threads.

The diameter of the scat can give you a general idea of the animal's size, helping you distinguish between, say, bobcat and cougar scat, or fox and coyote scat. On the other hand, a large bobcat and a small cougar, or a large fox and a small coyote, produce nearly identical scat, leaving you to look for more clues.

Scat Color

The color of scat is not as helpful a diagnostic tool as its size and shape. Meat eaters generally produce dark scat, but the scat of omnivores can vary greatly, depending on what they have recently eaten, which constantly changes with the seasons.

BE SAFE!

Animal scat may contain parasites and dangerous disease-causing organisms that can be transmitted to humans who handle the scat or inhale dry dusty particles of it. Raccoons, for instance, are the primary host of the roundworm *Baylisascaris procyonis*, which releases its eggs in raccoon scat. A human becomes infected by accidentally ingesting worm eggs after handling raccoon scat.

As much as possible, look rather than touch. If you want to break scat apart to examine the contents, wear disposable nitrile gloves and a dust mask, and use either a disposable stick or a tool you can readily disinfect outdoors.

To collect scat for future examination (preferably by a professional with experience in the safe handling of scat), turn a plastic zipper bag inside out and use it as a glove to enclose the scat. After sealing the collection bag, enclose it inside another clean plastic zipper bag for transport. Thoroughly wash your hands immediately afterward.

Mammal Scat Key

CYLINDER > 1" LONG	DESCRIPTION	SUSPECT
FLAT		
	twisted threads 1–1½" long	WEASEL FAMILY
NOT FLAT		
ONE END POINTED	large, cord-like, not segmented pieces, 4" long	COYOTE/ WOLF
	smooth, irregular pieces, 2" long	OPOSSUM
BOTH ENDS POINTED	ropy, connected pieces, 5–6" long	FOX
BOTH ENDS BLUNT	small, crumbly, irregular pieces, 1–2" long	SKUNK
	granular uniform cylinders, 2" long	RACCOON
	large segments, one end may be pointed, 4–5" long	CAT FAMILY

Scat's color also changes as the scat ages in the weather. Bobcat scat, for instance, can turn ashen white after only a few days of being dried by hot sun.

Scat that starts out white, on the other hand, comes from birds and reptiles. Anyone who has poultry already knows what bird poop looks like. Snake poop looks a lot like poultry poop but typically has a smaller white cap. The white cap consists of uric acid, the bird and reptile version of urine. Raptor droppings are also similar to poultry droppings but contain more white in proportion to the dark part, and they can be considerably looser — more like a splash of white paint.

Scat Odor

Mammals that eat large amounts of animal protein deposit scat with a strong odor. Although deliberately inhaling a scat's odor is a decidedly bad idea — because you could inhale seriously harmful organisms — the distinctive odors exuded by some scats can offer a clue as to the species that deposited it.

Cat-smelling scat comes, of course, from cats. The scat of bobcats and lynx smells like that of a house cat, only worse. A cougar's scat has the same acrid odor, only far worse. If you're having trouble deciding if a particular pile of poop came from a bobcat or a coyote, odor can be the deciding factor.

Coyote scat doesn't have a strong odor, although sometimes it can smell a bit musty. In any case, it doesn't smell much like that of a domestic dog.

The scat of a bear that's been eating meat smells rather foul, but unlike that of a cat or domestic dog. If the bear has been eating mostly berries, nuts, grass, and other vegetation, its scat will have a somewhat sweet, slightly fermented odor.

Fresh fox scat smells musty, a feature that can help you distinguish fox scat from coyote scat. Weasel scat smells musky, and sometimes skunk-like. Skunk scat can be decidedly foul smelling and, well, skunky. Raccoon scat has a distinctive pungent odor, but don't get too close, since raccoons carry a nasty roundworm parasite that can be transferred to a human via inhalation.

Contents and Consistency

Examining scat's contents can tell you whether the animal is a carnivore that eats primarily meat or an omnivore that has a more varied diet. Members of the cat family — bobcats, cougar, and lynx — cannot easily digest vegetable matter and therefore eat only meat. Because they digest meat so efficiently, their scat tends to be drier than that of most other predators. Minks, too, survive primarily on the flesh of other animals, as do birds of prey and snakes. Carnivorous predators — those whose survival relies on a diet of meat — have the highest motivation to get into your poultry yard.

Most predators are considered to be mesocarnivores, for which meat consists of at least half their diet. They are typically medium-size animals that are not shy about frequenting populated areas; sometimes, they are called urban mesocarnivores. They include coyotes, foxes, opossums, and raccoons. Scat from these predators may include hair and bone remnants, along with varying amounts of plant debris, especially fruit and berry seeds. When plant foods are plentiful, their scat may be somewhat softer than at other times. The presence of plant parts distinguishes their scat from that of cats. Mesocarnivores don't digest their food as well as cats do, and as a result their scat won't crumble as readily under pressure.

True omnivores are predators for which meat makes up less than 30 percent of their diet. Bears share this distinction with humans, and both species produce loose plops of scat when their diet consists primarily of juicy fruits and berries.

Dry or loose consistency is not a foolproof method of identification. Aging scat that has dried out will not be as soft as fresh scat. The scat of an animal suffering from constipation or diarrhea will have a different consistency from normal.

Scat Occurrence

Where and when you find scat can also offer important clues. In trying to identify a predator by its scat, take note of the following occurrences:

Time of day. Was the scat left during the day, or could it have been deposited overnight? This information may help you determine whether the predator is a night prowler or a day prowler.

Location. Is the scat on top of a rock, stump, or other elevated surface (indicative of fisher, mink, fox, or raccoon) or along a trail or where two trails cross (indicative of a bobcat or coyote)?

MAMMAL OR RAPTOR?

Mammal predators and raptors can sometimes leave much the same tragic scene. Both a coyote and a hawk, for example, may pick a bird's skeleton clean, leaving only the intact bones and a bunch of feathers. Here are some differences to look for:

1. A mammal leaves poultry feathers scattered randomly or trailing away. A raptor leaves a circle of plucked feathers.

2. Puncture holes made by talons are deeper than tooth punctures.

3. The tissue between a pair of puncture holes may be crushed by teeth but rarely is crushed by talons.

4. Multiple birds killed or missing are victims of a mammal. One bird killed or missing may be the work of a raptor.

5. Large mammals, such as bobcats and coyotes, prefer full-grown large poultry species, such as turkeys or geese. Few raptors bother them.

6. If you find a raptor feather dropped at the scene, it's a pretty good way to identify the perpetrator.

Frequency. Do you find the same type of scat at the same general spot on different days? How often? A coyote or other member of the dog family may return to mark the same spot with scat time and again.

One or multiple. Most species that prey on poultry create a latrine — a preferred place where an individual or family returns repeatedly to do business. An accumulation of scat in a latrine is a pretty good indication that the creator was not just passing through. Among predators that use a latrine are bears, bobcats, coyotes, raccoons, skunks, and weasels.

BIRD SIGNS

Predatory birds are of two varieties: those that eat poultry (raptors) and those that eat eggs (corvids and gulls). The raptors that most commonly prey on poultry are hawks and owls. Common corvids that love poultry eggs (and sometimes chicks) are crows and jays. Gulls also eat eggs and chicks.

Know Your Raptors

The word raptor derives from the Latin word *rapere,* meaning to seize. Otherwise known as birds of prey, raptors consist of a collection of genetically unrelated bird species that share in common four physical and behavioral features:

- Carnivorous diet, mostly consisting of freshly killed prey

- Binocular vision to help identify prey from a great distance

- Curved talons sharp enough to seize, pierce, and crush prey

- Hooked beaks strong enough to tear apart flesh

Raptors belong to three scientific orders: Strigiformes, Accipitriformes, and Falconiformes. Strigiformes are owls and are nocturnal (night hunters). All other raptors are either Accipitriformes or Falconiformes and are diurnal (daytime hunters).

Not all raptors are a threat to backyard poultry. Some have no interest. Others are too small or too weak to kill any but the tiniest chick. The main raptors found in the United States and Canada, and how likely they are to seek a meal in your poultry yard, are as follows:

Eagles are large and capable of carrying off a chicken. Although eagles rarely hunt poultry, I have lost full-grown New Hampshire chickens and pearl guineas to eagles. Of the two species in North America, bald eagles live near bodies of water; golden eagles live in mountainous areas.

Falcons eat birds, but the larger, more powerful species enjoy the challenge of catching prey in flight and don't see much fun of attacking poultry on the ground. The two smallest species — American

Raptors Most Likely to Be Interested in Poultry

ORDER	FAMILY	GENUS	SPECIES	COMMON NAME
Accipitriformes	Accipitridae	*Accipiter*	*A. cooperii*	Cooper's hawk
			A. gentilis	Northern goshawk
			A. striatus	Sharp-shinned hawk
		Buteo	*B. jamaicensis*	Red-tailed hawk
			B. lagopus	Rough-legged hawk
			B. lineatus	Red-shouldered hawk
			B. platypterus	Broad-winged hawk
			B. regalis	Ferruginous hawk
			B. swainsoni	Swainson's hawk
		Aquil	*A. chrysaetos*	Golden eagle
		Haliaeetus	*H. leucocephalus*	Bald eagle
Falconiformes	Falconidae	*Falco*	*F. columbarius*	Merlin
			F. mexicanus	Prairie falcon
			F. peregrinus	Peregrine falcon
			F. rusticolus	Gyrfalcon
			F. sparverius	American kestrel
Strigiformes	Strigidae	*Bubo*	*B. scandiacus*	Snowy owl
			B. virginianus	Great horned owl
		Strix	*S. nebulosa*	Great gray owl
			S. varia	Barred owl

kestrel and merlin — do occasionally visit poultry yards to carry off a small or young bird.

Harriers eat mainly rodents, reptiles, amphibians, and songbirds. The single species found in North America is called the northern harrier or marsh hawk. Harriers are sometimes called "good hawks" because, unlike most other hawks, they have no interest in poultry.

Hawks fall into two broad groups: fast-flying accipiters, which tend to live in forested areas and are often called chicken hawks after their predilection for poultry; and high-soaring buteos, which tend to live in open areas and are mostly interested in rodents. Hawks are the most likely raptor to hunt poultry during the daytime.

Kites prefer insects and small mammals. If you see a kite circling your poultry yard, it is probably hunting rodents. Kites of the five species found in the southern United States are not strong enough to do battle with a full-grown chicken, although one might occasionally nab a small chick of any barnyard species.

Osprey dine on fish and have no interest in poultry.

Owls are the oddballs of the bunch. They commonly hunt in the early morning and late evening, during periods of low light, but may occasionally hunt during the dark of night, as well as during full daylight. Owls eat mostly small mammals but do enjoy poultry when the opportunity offers.

Vultures, commonly called buzzards in North America, are scavengers that subsist largely on animals that have been dead for two to three days. If you spot a vulture in your poultry yard, it has most likely used its keen sense of smell to find something tasty that was already dead. They are included here because in flight they resemble eagles and on the ground they are sometimes mistaken for predators.

FOUR FASCINATING FACTS ABOUT RAPTORS

1. Raptors can see and hear up to eight times better than humans.

2. Their ears are concealed by feathers, with one ear higher than the other, to accurately pinpoint a source of sound.

3. A hawk can carry prey weighing no more than about half its body weight; an owl can carry up to three times its own weight.

4. The two greatest raptor threats to poultry are the nocturnal great horned owl and the diurnal red-tailed hawk.

Raptors in Flight

Members of the five groups of raptors that are most likely to fly over a poultry yard may be identified by shape and size, as revealed by their silhouettes against the sky, and by their somewhat unique flight patterns.

Eagle: large bird; noticeably large head and beak; fanned tail; long, broad wings that require a lot of energy to beat. The eagle therefore spends little time flapping and most of its sky time soaring and gliding with its wings held flat.

Falcon: small bird; roundish head; short parallel tail, fanned when hovering; short pointed wings. In flight, the falcon flaps its wings almost continuously. It is the aerial acrobat of the bunch.

Hawk, accipiter: midsize bird; long tail; short, rounded wings. The flight pattern consists of several snappy wing beats alternating with a short glide. Sometimes, though, an accipiter will soar like a buteo.

Hawk, buteo: midsize bird; short, fanned tail; long, broad, rounded wings. The buteo hawk glides effortlessly on air currents for long stretches without flapping its wings; infrequent wing beats appear labored. While the buteo typically soars high above open fields, it may also fly low or even hover when it detects prey.

Owl: large bird; large, round head; short, wide tail; wide wings that allow it to glide more than flap. Owls are rarely seen flying during daylight.

Vulture: large bird; small head; long parallel tail (turkey vulture)/short tail (black vulture); long, wide wings. Turkey vultures rarely flap their wings; black vultures alternate rapid, deep wing beats with a short glide.

Know Your Corvids

Members of the crow family — including ravens, magpies, and jays — are among the world's most intelligent birds. Collectively known as corvids (all being members of the family Corvidae), their brain-to-body ratio is only slightly smaller than a human's.

Thanks to their high intelligence, corvids have a broad vocabulary of sounds and sometimes imitate sounds made by other birds, humans, and even machinery. They also have excellent memories and can learn from fellow corvids, especially when it comes to finding tasty things to eat — such as poultry eggs laid outdoors (or even indoors) and sometimes hatchlings.

Crows and ravens (*Corvus* spp.) are found throughout North America. They are rather large, robust birds with stout legs and strong, grasping toes. Their plumage is a glossy black with a metallic sheen. Magpies (*Pica* spp.) are distinctive from crows and ravens in being smaller, having longer tails, and having black and white plumage instead of being all black. Jays (*Cyanocitta* spp.) are the most colorful members of the crow family, but also the most reclusive. On our farm we hear their brash call more often than we see the birds themselves.

The larger corvids may carry eggs away from the nest to eat elsewhere, so you may or may not find shells in or near the nest with jagged holes pecked into the sides. Jays, being smaller than other corvids, are more likely to leave shells in the nest, rather than carry them away.

Raven   American crow   Magpie   Blue jay

FOUR FASCINATING FACTS ABOUT CORVIDS

1. Crows in the wild can recognize individual humans, but few humans can recognize individual crows.

2. Nearly all corvid species have been seen using tools to acquire food.

3. When an egg is too large for a corvid to carry in its mouth, it pierces the shell and carries the egg by sticking its beak through the hole.

4. Corvids will mob and attack any person or animal, including other corvids, that they find threatening.

Outwitting Jays

Dave Holderread, Oregon, Holderread Waterfowl Farm

It was morning here on the waterfowl farm and time to gather eggs. Spring was in the air. Listening carefully, you might detect the chatter of hungry songbird nestlings begging for their next meal as a parent arrives back at the nest with a beakful of groceries.

Moving from pen to pen, I gathered duck and goose eggs from ground-level nests bedded with clean straw and wood shavings. Before placing each egg in the egg carrier, I labeled it for breed and pen number.

Entering one of the outdoor pens, I caught a glimpse of a stealthy jay flying out of the duck shelter. And there, in one of the nests, it had left its tell-tale visitation signature: an intact duck egg with a hole about the diameter of a pencil pecked in the upper surface of the shell.

In our experience, jays pilfer eggs mostly when they are feeding their nestlings. They will sometimes carry off small eggs of bantam breeds, but jays typically puncture the shells of normal-sized chicken and duck eggs and then carry some of the albumin and yolk to their nestlings, leaving the shell and much of its contents in the poultry nest. We have lost few goose eggs to jays over the decades, because of the hardness of the shells and because geese carefully cover their eggs after each visit to the nest.

We have been able to virtually eliminate duck egg losses to jays by enclosing our laying flocks from dusk until about 8:00 a.m. in night quarters with no openings larger than 1½ by 1½ inches — too small for a jay to slip through. Domestic ducks typically lay the majority of their eggs before 8:00 a.m. As soon as we let the ducks into their daytime quarters each morning, we gather their eggs before the jays have time to discover them.

Corvids and Gulls Likely to Eat Poultry Eggs

ORDER	FAMILY> SUBFAMILY	GENUS	SPECIES	COMMON NAME
Passeriformes	Corvidae> Corvinae	Corvus	C. brachyrhynchos	American crow
			C. ossifragus	Fish crow
			C. caurinus	Northwestern crow
			C. corax	Common raven
			C. cryptoleucus	Chihuahuan raven
		Pica	P. hudsonia	Black-billed magpie
			P. nuttalli	Yellow-billed magpie
		Cyanocitta	C. cristata	Blue jay
			C. stelleri	Steller's jay
Charadriiformes	Laridae	Larus	Larus spp.	Gulls

Know Your Gulls

North America has some two dozen gull species of the *Larus* genus, all of which are medium to large birds with gray or white feathers, long, hooked bills, and webbed feet. Colloquially known as seagulls, gulls live mainly along the coastlines of oceans, bays, and large lakes, although some either live inland permanently or may fly inland seeking food.

Gulls are scavengers, omnivores that will eat almost anything edible. Larger species are particularly fond of eggs and hatchlings, including those of backyard poultry. Like corvids, gulls may carry eggs away to eat elsewhere or may eat them in the nest, leaving shells with longwise tapered slots pecked into the sides.

Bird Tracks

If you are trying to determine whether a predator is, say, an owl or a raccoon, or a crow or a ground squirrel, finding tracks would be helpful. Bird tracks are markedly different from mammal tracks. The trick here is to distinguish between tracks made by your poultry from those made by predatory birds visiting your poultry yard. Bird tracks are identified according to the number of visible toes, whether the toes are pointing forward or back, and whether or not the bird's feet are webbed. Recognizing the following four track patterns left by birds can be helpful in making an identification.

Classic. Most birds' toes are arranged in the classic pattern, technically known as anisodactyly (from the Greek words *anisos*, meaning not equal, and *dactylos*, meaning toe). These tracks show one toe pointing backward and three pointing forward. Birds that leave classic tracks include diurnal raptors (eagles, falcons, and hawks), corvids (crows, jays, and ravens), and pigeons.

Game bird. Also technically anisodactyl, game bird tracks differ from classic tracks in that the back toe either appears shorter or does not show at all. This track is made by gallinaceous birds (those related to chickens), including guineas, turkeys, pheasants, quail, and, of course, chickens.

Webbed. These tracks, also technically classified as anisodactyl, are similar to game bird tracks except they show webbing between the forward-facing toes. Among predators, egg-eating gulls make webbed tracks. So do ducks and geese.

K-shape. This pattern, technically known as zygodactyly (from the Greek word *zygotos*, meaning yoked or paired), shows two toes facing forward, one to

Measuring Bird Tracks

Measuring the size of a track can give you a clue as to a bird's size, and therefore its identity. When measuring a game bird or webbed track, with the back toe short or missing, the claw is included but the back toe (hallux) is not. This measuring method allows for a more reliable comparison between similar species.

Classic (anisodactyl) tracks. Measure length from the tip of the claw of the back toe to the tip of the claw on the middle front toe. Measure width from the tips of the claws on the two outside front toes.

Game bird tracks. Ignoring any visible mark left by the back toe, measure length from the base of the foot pad to the tip of the claw on the middle front toe. Measure width from the tips of the claws on the two outside front toes.

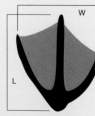

Webbed tracks. Measure length and width the same as for game bird tracks.

K-shaped (zygodactyl) tracks. Measure length from the tip of the claw on the back toe to the tip of the nearest front toe (forming a straight line). Measure width from the tip of the claw on the sideways-pointing toe across the widest part of the track.

the side, and one facing back, forming a K. This is the track pattern left by owls, among other zygodactyls (most of which leave an X-shaped track instead of a K). The owl's right foot prints a typical K; the left foot prints a reverse ⋊.

Bird Track Patterns

Birds make four kinds of trail patterns. As with mammals, along with track size and shape, the straddle and stride of a bird's trail pattern can offer clues as to the bird's identity. Stride is measured from the tip of the claw on the middle front toe of one track to the tip of the claw on the middle front toe of the next track. Straddle is the distance between the outer toe of the left track and the outer toe of the right track, with care taken to measure straight across, and not at an angle.

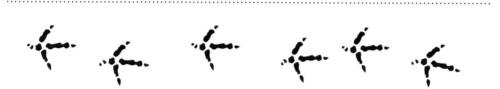

Walking. The walking trail pattern is characteristic of birds that spend most of their time on the ground, which includes all the species that produce tracks in the game bird and webbed groups. All backyard poultry species walk. Raptors, crows, magpies, and ravens may also walk. Walking tracks are closely spaced, with one in front of the other to form a continuous trail that may be either a straight line (such as the trail made by a chicken or turkey, by putting one foot directly in front of the other) or zigzaggy (such as a trail made by a duck or a goose, by waddling).

Running. The running trail pattern is similar to the walking pattern, and in fact often starts as a walk. The main difference is that the tracks are farther apart; the stride can be five times or more the length of one track. While a walker always has one foot on the ground, a runner is briefly airborne between steps. In fact, a running trail may disappear abruptly where the bird took flight.

Hopping. In the hopping trail pattern, both tracks appear side by side or nearly so. This pattern is made by birds that spend most of their time in trees. It is the usual pattern of songbirds, including jays, but may also be made by crows, magpies, and ravens. Stride is measured from the tip of the middle front claw on the farther forward foot in one set of tracks to the middle front claw on the farther forward foot in the next set (which may or may not be on the same side).

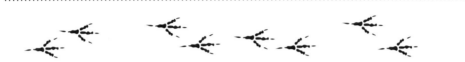

Skipping. The skipping pattern is similar to the hopping pattern but is generally made by a bird moving faster than a hop, so one foot lands ahead of the other. This pattern is made by birds that spend time both in trees and on the ground, including crows, magpies, and ravens. Paired tracks are farther apart than in a hopping pattern and may alternate with walking.

Cough Pellets

Wild birds have the same two-part stomach as backyard poultry. In the first part, or true stomach, feedstuffs are partially digested by acids and enzymes. The second part, or gizzard, pulverizes the more fibrous particles of the partially digested food. What happens next is where the digestive process of domestic poultry differs from that of wild birds that routinely feed on items with indigestible parts.

Where the poultry gizzard passes the pulverized food along to the small intestine, the gizzard of a raptor, corvid, or gull squeezes out the pulpy, acidic digestible slurry (called chyme) and passes it to the small intestine. The remaining dry, undigested matter — consisting of bones, feathers, hair, insect parts, or other indigestibles — is sent back up to the true stomach, where it is stored until the bird is ready for its next meal.

Before the bird can eat again, it must get rid of the undigested pellet. It does so by coughing up the dry pellet (a process called casting), similar to a cat coughing up a hairball. If you find a cough pellet, its size, shape, and contents offer a clue as to the bird's identity.

Since pellet size is directly related to the shape and size of the bird's gizzard, the larger the bird, the larger the pellet. Owl pellets, for example, range from 1 to

A cough pellet may be carefully teased apart to reveal what the bird has been eating, offering a clue as to the bird's identity. Please note, however, that a pellet may contain bacteria that can cause illness, so avoid handling a pellet with your bare hands. Instead, wear disposable gloves while examining the contents of a pellet, and carefully wash your hands afterward.

An owl gulps down what it eats — feathers, bones, and all — sometimes even swallowing a whole chick or other small animal. The owl's rather weak digestive juices aren't acidic enough to dissolve bones. Compared to the pellets of other birds, the owl's pellet will therefore contain more bones, and some of the bones may be intact. A large pellet containing lots of bones may be easily identified as that of a great horned owl.

An owl, in fact, has such an inefficient digestive system that nearly everything it eats may be identified in its pellet. If, for example, you find a pellet that's filled with insect parts, the owl that left it wasn't after your poultry but more likely dined on insects attracted by your nighttime security light.

An eagle, falcon, or hawk, by contrast, plucks its prey and then uses its beak and feet to tear the flesh off the bone and swallow a bite at a time. And the bird's digestive acids are strong enough to dissolve much of what the bird eats, includ-

ing some of the bones. A pellet left by one of these diurnal raptors contains far fewer bones than that of a great horned or barred owl, and any bones in the pellet are likely to be broken into pieces. Overall, the pellet left by an eagle or hawk reveals less about what the bird has been eating than does an owl's pellet.

A vulture's rather large pellet looks similar to that of other diurnal raptors. One clue that tells you it was left by a vulture is the presence of plant matter the bird may have incidentally swallowed while feeding on carrion.

Cough pellet cast by a red-tailed hawk

4 inches long. A great horned owl's pellet is the longest and thickest. The smaller barred owl pellet, on the other hand, is similar in size to that of a falcon or hawk.

A pellet has a smooth surface and is either cylindrical or round, although a gull's roundish pellets tend to have an irregular shape. A cylindrical pellet can look quite a bit like mammal scat. But you can easily tell the difference. For starters, a cough pellet has no odor. Further, pellets are expelled individually, rather than being connected in segments.

A raptor's pellets are denser than most mammal scat, and therefore more difficult to break apart. Corvid pellets, on the other hand, readily fall apart because, unlike raptor pellets, they lack hair to provide cohesion. You are therefore more likely to find the denser pellet of a raptor than the more crumbly pellet of a corvid or gull.

A freshly expelled pellet glistens with a coating of iridescent mucus that has smoothed the pellet's way past the bird's throat. An older pellet looks dry, unless it has been moistened by rain. Wet weather causes a pellet to deteriorate. The smaller the pellet, the faster it deteriorates.

Pellets are commonly found beneath the spot where a bird roosts. An owl, however, typically drops a pellet at the site where it is hunting. If you find a pellet in your poultry yard, it was most likely left by an owl. You are less likely to find pellets in your yard left by diurnal hunters because they typically don't hang around where they easily could be detected. Hawk pellets more often accumulate beneath the bird's perching or nesting site.

Bird Poop

While the indigestible matter goes back to the bird's stomach to be coughed up, the chyme continues to the small intestine, where the nutrients are absorbed, and then to the large intestine, where much of the moisture is absorbed. Finally, liquid and solid waste enter the cloaca to be expelled.

Quite unlike the semisolid poop of backyard poultry, which results from a diet of mostly plant matter and insects, the poop of carnivorous and scavenging birds consists mostly of a uric acid — a bird's equivalent of human urine. The resulting poop is mostly a white fluid that is often compared to splattered whitewash. Unlike with mammals, the species of a predatory bird is therefore difficult to identify from its scat.

Still, the appearance of bird poop may offer some clues. Raptors quite often leave poop while they are enjoying a meal of poultry. An owl poops straight down, leaving a chalky white heap or puddle. A hawk or falcon raises its tail and sprays, sometimes for a distance of 3 feet or more, creating a white streak or splash that radiates away from the circle of plucked poultry feathers. A corvid that has been eating mostly plant matter, on the other hand, leaves droppings that are virtually indistinguishable from a chicken's.

4. FOILING THE PERPS

> **"It is better to learn wisdom late than never to learn it at all."**
>
> Sherlock Holmes

Predators tend to seek food that's easy to get. The way to prevent predation, therefore, is to make your poultry more difficult to get than the predator's natural fare — such as rodents or carrion — in the surrounding environment. By far the two best methods for making your poultry hard to get are to lock them indoors at night and to install a sound fence around their day yard. Because the subject of fencing is so important, it has its own chapter, to follow. Here we'll examine additional ways to keep predators at bay.

ELIMINATE POINTS OF ENTRY
(BEST PLAN)

Once a predator gets a taste of poultry, it's likely to come back for more. The first line of defense is to make sure it doesn't get that first taste. In the event a predator does breach your security, the obvious course of action is to take immediate measures to prevent further losses. The following defensive measures should become part of your poultry predator prevention program, if they aren't already.

Pophole closer. The easiest way to eliminate point of entry is to confine poultry inside, if not all the time, at least at night. An electronically operated pophole door that automatically closes when the sun goes down and opens at sunup is helpful if you aren't always Johnny-on-the-spot to lock up your chickens at dusk and let them out at dawn.

Leaving the pophole open after dark or opening it in the early morning hours is an invitation to owls. If your predator is a coyote, fox, or raccoon, however, you'll need to keep your chickens in until late morning, by which time these critters usually have called it a night. Some automatic doors have, in addition to a daylight sensor, a timer override that lets you set the time you want the door to open and close. Most automatic doors are battery operated, or may be converted easily to battery or solar where electricity is not readily at hand.

Seal off openings. Cover all windows and ventilation openings with ½-inch hardware cloth — not chicken wire, which a fox, dog, raccoon, or bear can readily rip. Check under the roof along the eaves for less visible openings that need to be sealed off as well.

Solid coop floor. Although well-finished concrete is the most expensive type of coop flooring, it requires minimal repair and upkeep, is easy to clean, and keeps predators from tunneling into the coop. Alternatively, a deep concrete perimeter foundation will discourage burrowing predators. If the coop has a dirt floor, secure hardware cloth or other sturdy mesh beneath a layer of soil to deter rats, foxes, weasels, and other burrowers.

Elevated coop. If your coop floor is raised off the ground with an air space underneath, make it at least 1 foot high to discourage rats, snakes, and weasels from taking up residence underneath and using that vantage point to get into the coop. Elevating the coop another foot or two will discourage most other predator break-ins. A small coop may be elevated easily using stout corner posts. A large coop could be raised on piers, but a better option for a large coop is a solid perimeter foundation of either blocks or concrete.

Secure coop door. If the people-size door to your coop is not enclosed within a fenced yard, make sure it shuts tight and can be securely latched or locked. Install latches high enough that curious toddlers can't open them and forget to reclose them.

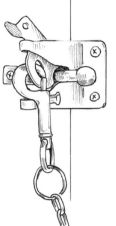

◄ Aside from humans, most predators can't open a double-action latch.

▲ A conical tree guard will keep predators from climbing a tree to drop down into your poultry yard.

Double-action latches. Latch all doors and gates with devices requiring two different motions, such as lifting and pulling. Further secure each latch with a snap hook. Secure the hook to the gate with a length of chain so you can't accidentally drop the hook and lose it or stick it in your pocket and forget to slip it back on the latch.

No cover vegetation. The harder a predator has to work to get through the fence and into your poultry yard, the less likely it is to try, especially if the area outside the fence leaves the animal exposed and visible. Shrubs and tall grass on the outside of the yard may provide attractive landscaping, but they encourage predation by providing handy cover.

Tree guard. If trees are growing close to the outside of the fence, raccoons, gray foxes, ground squirrels, and other mammalian predators may climb up into the limbs and then drop down into the poultry yard. A conical guard made of aluminum flashing or sheet metal will keep climbers from gaining entry to your run by climbing up a tree, pole, or post. Fasten the cone together with wire or screws and paint it to blend in with your landscape. Such a predator guard on a live tree needs to be adjusted periodically to accommodate the tree's ever-increasing diameter.

Overhead netting. Some of the most difficult predators to control are those that come down from the sky. If your poultry yard is small enough, you might cover the top with aviary netting or similar material. In an area where raptors are dense or aggressive, or where raccoons and other climbers may gain entry by climbing on top of the pen, use welded wire or other sturdy mesh strong enough to support the predator's weight.

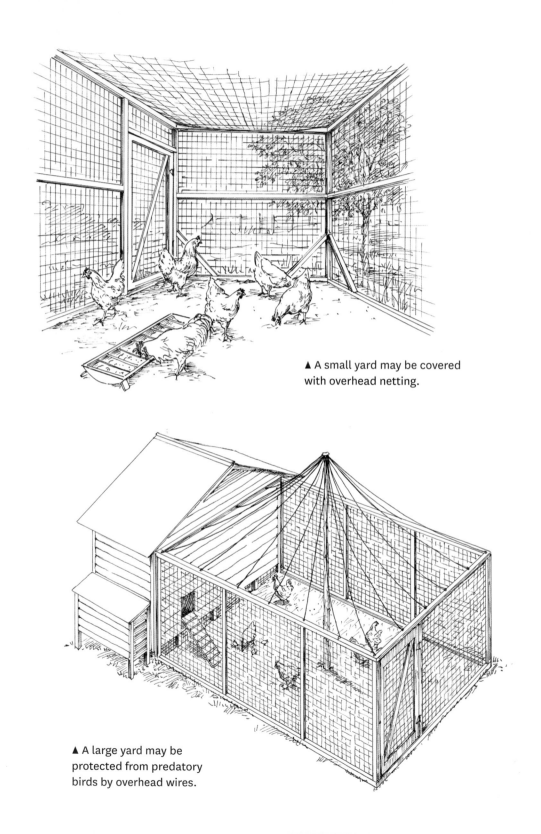

▲ A small yard may be covered with overhead netting.

▲ A large yard may be protected from predatory birds by overhead wires.

A poultry yard that's too large to easily or economically cover might be crisscrossed with ropes or wires strung no more than 2 feet apart. Most raptors, corvids, and gulls won't fly into a space that looks like it might be difficult to get back out of. Whether you are using netting or strand wire, make the cover 7 feet off the ground, or at least high enough not to snag your head and any equipment you might use to maintain your poultry yard.

Portable pen. If you use a portable pen, moving it every couple of days will confuse predators, or at least make them suspicious enough to be cautious about approaching. If the pen doesn't have a solid roof all the way across, cover the open area with overhead netting, and add a second layer 6 inches above that.

Hardware cloth or 1-by-2-inch welded wire will keep raccoons from reaching through to grab a bird.

Anchor the pen with skirting that's tightly woven and close to the ground. Each time you move the shelter, double-check for dips where weasels might weasel in. As with a stationary yard, keep grass and weeds well mowed, since many four-legged marauders prefer not to expose themselves by crossing an open field. Avoid placing the pen within 100 yards of a tree or other possible raptor perch site. Create a secure foraging area around the portable shelter using electrified plastic netting (described on page 110) energized by a solar- or battery-powered energizer.

COOPING WATERFOWL

Many poultry flocks include ducks as well as chickens, and some poultry keepers train their ducks to go into the coop at night with the chickens. However, unlike chickens, ducks don't sleep the night through, and their restlessness can disturb the chickens' sleep. A better plan is to coop waterfowl separately from chickens and other land fowl.

Waterfowl tend to have an aversion to being cooped up all night. However, if you start training them while they're young, you should have no trouble getting ducks or geese to go inside a protective enclosure at night. Even mature waterfowl can be trained by feeding them in the evening at the enclosure, near enough to easily herd them inside.

Because ducks don't sleep the night through, they can be disturbed easily by passing lights coming through the shelter window, causing them to panic and trample one another. Keep this factor in mind when orienting windows in a coop intended for ducks.

Perils of Free Ranging

Bob Bennett, Vermont, rabbit keeper and author, *Storey's Guide to Raising Rabbits*

I don't have chickens now, but my family had them for several years when I was a kid. We kept them in a coop with a covered run. I live in a semi-rural area, on 10 acres or so. So does the neighbor across the road. It's a dead-end street with 12 houses on it, each averaging about 10 acres. Behind me are dairy farms, separated by some houses and many acres of woodland.

We have lots of deer, foxes, turkeys, bobcats, coyotes, coons, woodchucks, 'possums, and rabbits of course, and who knows what else, all beneath bald eagles, red hawks, turkey vultures, geese, ducks, herons, and who knows what else, as we are close to the large Lake Champlain.

Some people don't approve of keeping chickens in a covered run, including the surgeon who lives across the road.

He keeps them in a coop at night, but they run around his several acres during the day. I should put that in past tense because he no longer has chickens. They got picked off one by one by foxes, bobcats, and who knows what else.

He called them free-range chickens and let them out to be 'humane.' I called them free-lunch chickens until I found eggshells on my front yard, perhaps delivered by a raccoon. Then I called them free-breakfast chickens.

Now my surgeon neighbor calls them history and gets his eggs at the supermarket.

DISCOURAGE PREDATORS
(GOOD IDEA)

With your coop and poultry yard secure, you can still do a few things to discourage predators, or at least avoid encouraging their presence. One is to remove anything that might potentially attract wildlife. Another is to install some of the many available protective devices designed to frighten predators way. And finally, you might consider employing a guardian animal to protect your flock.

Eliminate Attractants

Wild animals are constantly looking for food, water, and shelter. Intentional or unintentional sources serve as attractants that encourage predators to stick around and routinely seek easy handouts. As they become bolder and less fearful of humans, they also become more likely to start trying to figure out how to get to your poultry. Further, the ready availability of handouts can encourage increased reproduction, resulting in a greater number of predators seeking poultry snacks. Eliminating the following wildlife attractants creates a less hospitable environment, encouraging predators to move on to where conditions might be more to their liking.

Garbage. Store garbage cans inside a garage or enclosed shed, or secure metal containers so they can't be tipped over and tie down lids with sturdy straps. Readily available garbage attracts bears, foxes, opossums, raccoons, skunks, and other types of wildlife that relish carrion as well as live poultry.

Outdoor grill. If you use an outdoor grill, clean it thoroughly after each use to ensure that residual odors don't attract bears and other wildlife. Better yet, store a cleaned grill and any other outdoor cooker in a securely enclosed garage or other outbuilding.

Poultry feed. Store feed in sturdy containers with tight-fitting lids, preferably inside a securely enclosed building to minimize the attractive odor. Promptly clean up any spilled feed.

Pet food. If you feed poultry, dogs, or cats outdoors, depending on the type of feeder you use, close it tight at night or bring it indoors where wildlife can't gain access. Opossums and raccoons are especially attracted to pet food.

Wildlife feeders. In some states, feeding wildlife is illegal. Even where it's not, don't attract predators by leaving out leftovers or otherwise feeding wildlife. A deer feeder, for example, may attract a cougar seeking a meal of venison. If you have a bird feeder, use a style that prevents seed from accumulating on the ground, and bring it indoors at night to prevent recurring visits from bears or raccoons.

Drinking water. Drinkers for poultry or pets left outdoors overnight can be powerful attractants, especially in a dry area or during times of drought. Rats, for one, won't hang around long where drinking water is not consistently available.

Fallen fruit. Clean up fruit that falls from orchard trees. Not only can it attract coyotes, opossums, raccoons, and other wildlife, but if it remains on the ground long enough to ferment, animals that eat it may become tipsy and unpredictable.

Hiding places. Eliminate nearby dense vegetation and remove or secure firewood piles and similar places where wildlife may seek refuge, raise young, or hide and watch for an opportunity to gain entry to your poultry yard. Likewise, seal off access beneath the coop and around nearby building foundations and porches.

Landscaping the Run

Vegetation growing inside the poultry run can be both beneficial and harmful. Low-growing shrubs give poultry a place to hide in the event they feel threatened. However, if a predator gets inside the yard, it will chase them into their hideout. Unfortunately, once a bird feels safely hidden, it stops running and hunkers down, becoming easy prey.

Trees inside the run can also be beneficial or harmful. They provide a shady place for birds to rest on hot days and protect poultry from the types of raptors that avoid hunting beneath trees. Owls and accipiter hawks, on the other hand, typically land in a tree to do a fast surveillance before swooping down on an unsuspecting victim. However, when no trees are available, they are just as happy to use a handy fence post or any other high point, including a rooftop or utility pole, within 100 yards of the poultry yard.

Bottom line: You'll have to decide for yourself if the benefits to your poultry of low-growing vegetation and shade trees outweigh the potential risks of predation.

Protective Devices

Aside from a well-built fence, no single protective device provides a foolproof way to deter all comers. Most of the suggested protective devices will work for a while, giving you time to beef up your poultry yard security. Unfortunately, many such devices are little more than expensive toys that can be time-consuming to maintain and will eventually fail against any determined and persistent predator, especially one that has already tasted success and comes back for more. The hungrier a predator is, the quicker it will lose its fear of any scare tactic you might devise.

My poultry yard is protected by a combination of LED lights that flash in the dark, motion-activated security lights, and wind-catching DVDs. I'd like to say that they have successfully deterred all predators, but that wouldn't be true. We still occasionally lose a chicken or guinea, almost always because an LED light stopped flashing or a storm blew down the DVDs. The message here is that all of these devices need frequent monitoring and occasional replacement.

Flashing LED lights. Blinking red LED lights are one deterrent that, when properly installed, can work for nearly any predator prowling or flying between dusk and dawn. Among the many solar-powered devices designed for this purpose, some have one blinking light and some have two. I've used both and they seem to work equally well. The constantly flashing pulse of light gives wary predators the impression that they are being watched, causing them to skulk away to avoid a potentially hazardous confrontation. The technique even works on two-legged thieves, who believe the coop is wired with a security system. Top brands include Aspectek Predator Eye Pro (*www.aspectek.com*), Nite Eyes (*www.interplexsolar.net/solar-night-eyes*), Nite Guard (*www.niteguard.com*), and Predator Guard (*www.predatorguard.com*).

We had a consistent problem with early-dawn pilfering of feed, eggs, and chickens that stopped abruptly after we mounted one of these devices next to the automatic pophole door. The trick to effective installation is to mount the unit at eye height for the predator in question, and mount enough units that a predator will encounter one straight-on no matter which direction it approaches from.

Security lights. Some predators, especially bobcats and cougars, may be discouraged by lights left on at night. Even if security lights don't deter predators, they will help you see more quickly should you have to rush out in the dark to identify the cause of a commotion at the coop. Remember, though, that just like us humans, chickens need undisturbed sleep, so make sure the security light doesn't brighten the inside of the coop. And that it doesn't interfere with a light-sensitive pophole door opener.

Pole shockers. An electric pole shocker is designed to shock a raptor that lands on top of a wooden fence post or other perch or pole, discouraging the

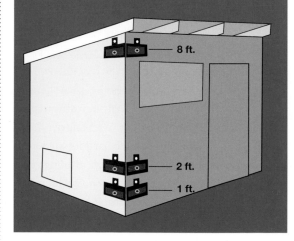

BEST HEIGHT FOR FLASHING LED LIGHTS

For small predators: 1 foot off the ground
For large predators: 2 feet off the ground
For flying predators: 8 feet off the ground

predator from hanging around. However, for this system to work, all potential perching sites must be protected or eliminated. Since raptors generally favor areas with trees, and since citizens aren't allowed to add protective devices to utility poles, for most situations this frequently suggested method is eminently impractical.

Wind catchers. Flashing devices that catch light as they flutter or spin in a breeze (including a breeze generated by a raptor's wings) can be an effective deterrent. Such things include old CDs or DVDs, Mylar predator repellent tape, and shiny metal objects like disposable aluminum pie tins. Fluttering cloth streamers, such as might be made from strips of old bedsheets or survey tape, also work, as do windsocks and pinwheel wind spinners. The problem with most of these devices is that they work best in wind, and persistently windy weather will eventually work them loose or break them, so that they require frequent replacement, as well as leaving bits of debris in the yard for you to pick up. Like flashing LED lights, wind catchers work best at the predator's eye level, high up for raptors, closer to the ground for mammals.

Fly curtains. Where corvids enter an open pophole to gain access to eggs inside the coop during the day, chicken keepers have had success keeping them out by hanging colorful fly curtain strips in the pophole opening. We hang overlapping PVC strips in our barn doorways and haven't had any crows get in, although our chickens and guineas have no problem going through them. I think corvids might be reluctant to enter if they can't fully see whether or not it's safe inside.

Fladry. This deterrent is a specific type of wind catcher consisting of red or orange pennants spaced about 18 inches apart on a strong string, wire, or rope. Its intent is to discourage cautious wildlife, particularly wolves. It is sometimes used in conjunction with a fence, and sometimes in place of a fence as a psychological barrier (which I wouldn't count on to protect chickens from wolves or anything else). A fladry line works only when air movement causes the pennants to flutter (the strange word *fladry* derives from the German word *flattern*, meaning to flap or flutter), and even then it only works until predators decide your birds are more tempting than the pennants are scary. Once a predator crosses the fladry line, it completely loses its fear of it. Hanging pennants from an electrified wire makes the fladry line marginally more effective.

Motion detectors. Motion-sensitive lights or sirens that go on suddenly can startle a skulking predator into taking cover. Such alarms may be just as easily set off by, and disturb, you or your own pets or poultry. Further, unless the predator happens to be just passing through, a smart and determined animal will eventually realize that the sudden flurry of lights and/or sound is all bark and no bite, at which time the deterrent will become completely ineffective. Further, if you use motion-sensitive lights and an automatic pophole door opener, make darn sure the light doesn't shine on the door opener's light sensor or you will inadvertently invite the predator to a banquet.

A motion-activated water sprayer that blasts an approaching predator with a sudden burst of water works against some predators, especially those of the cat or dog persuasion. The downsides to this method are that it requires available running water, and a persistent predator may soon find an alternate route to your coop to avoid getting soaked.

Noise makers. Playing a radio all night, set loudly to a talk show network (or fitted with a motion detector that turns the radio on when a predator approaches), is a popular scare tactic. Wind chimes also work to a certain extent, at least until prowling predators get used to the musical sound. Old-timers once used explosive devices known as bird bombs or shell crackers for predator control, but they never proved to be particularly effective, and today such pyrotechnic devices require a permit from the United States Bureau of Alcohol, Tobacco, Firearms, and Explosives (ATF).

Odorous fumes. Posting predator urine (*www.predatorpee.com*) around the perimeter of your run is costly and time-consuming but can provide a temporary measure against four-legged predators. For details, see page 97. Ammonia-soaked rags also work against most predators, although some may think it's a urine mark and cover it with their own urine scent, which can also happen with predator urine. Hanging sweaty or otherwise soiled clothing with a strong human odor can cause predators to temporarily steer clear. All such odorous fumes need replacement every two to four weeks, or more often in rainy weather.

Any number of ready-made or recipes for homemade smelly concoctions designed to repel wildlife are available online, some of which may repel one type of predator but attract another, possibly worse, predator. For example, citronella may repel feral cats but attract bears. Pepper spray may repel opossums and skunks but may attract grizzly bears.

Mothballs. Forget the ubiquitous pesticides known as mothballs. They may deter opossums, raccoons, skunks, and snakes, but when they are eaten or their fumes inhaled, they are toxic — especially to children, pets, and poultry.

Scarecrows, fake owls, bird balloons. Save your money. These items may work, but not for long.

Alternating tactics. The effectiveness of any of these scare devices may be prolonged by frequently moving them around and alternating their use. Predators avoid potentially dangerous situations. By frequently altering your scare tactics, you create a situation that predators have difficulty understanding, causing them to steer clear.

A scarecrow left in one place for long will scare hawks about as much as it scares these chickens.

Poultry Guardians

The natural defender of a chicken flock is a rooster. For geese, it's the gander. And so forth for other poultry species. The male of the bunch will generally sound an alarm when he senses danger, warning the rest of the flock to run and hide. He may or may not fend off a predator, and indeed may himself fall victim to a predator, which is perhaps one reason why so many more poultry males hatch than are needed for social cohesion and procreation.

Turkeys of both genders can be pretty defensive against smaller predators, especially birds — such as crows, jays, or magpies — that visit the yard to pilfer eggs. Guinea fowl are especially notorious for sounding a predator alarm. If the guinea flock is large enough, it will circle an intruder and drive it off. If the intruder is a small creature, such as a rat or a snake, the guineas may kill it. On the other hand, a guinea can fall victim to a venomous snake defending itself from attack.

To protect the entire flock, especially free-ranging poultry, a well-trained guardian dog is a good investment. A guardian dog can be effective against coyotes that hunt singly or in pairs, but less so against wolves or coyotes that hunt in packs. If you opt for a canine, be sure he's reliable. Running, flapping, squawking birds are a mighty big temptation for a dog, which may merely have playful intent but end up chasing or shaking birds to death. Many chicken keepers have lost birds to their own happy-go-lucky family dog that got overly frisky or just plain bored. Indeed, a colleague of mine once remarked that her thriving poultry business came largely from repeat customers, thanks to domestic dogs — not just neighborhood dogs and roaming packs, but also pets owned by the same people who time and again purchased flock replacements.

The best way to train a dog to be poultry savvy is to put it on a leash and have it accompany you when you visit your flock. Train the dog to sit and watch while you work. As time goes by, the dog will become less excited by the presence of birds and begin to express watchful curiosity. When you feel that the dog has been sufficiently habituated to your flock, let it off the leash under supervision. Until you are completely certain that your birds are safe with the dog, any time you are not around to supervise, keep the dog in an adjoining yard where dog and flock can watch one another and get used to each other's habits.

Like everything in life, this method is not foolproof. My neighbor kept a dog, a flock of chickens, and a garden within a fenced backyard. The dog kept the chickens out of the garden by herding them away whenever they wandered to that end of the yard. One of the hens hatched a brood of chicks, and the dog let the chicks cuddle around his head while he slept. One day all the chicks suddenly disappeared. My neighbor eventually found them buried in the garden, apparently killed by the dog, perhaps in play. That dog never bothered grown chickens in its own yard, but one day when it got out and crossed the road to my yard, it maimed several of my chickens. So if you think *my dog would never do that,* you might want to reconsider.

A well-trained and reliable dog can make a good poultry protector.

Most predators are afraid of a barking dog and will steer clear. The dog doesn't necessarily have to be in the same yard with the chickens, only near enough to bark when it detects an intruder. An alternative is to let the dog guard the poultry yard at night, while the chickens are safely enclosed within their coop.

Ironically, some poultry predators actually help protect your flock. Crows and jays, for instance, will gang up on hawks, a behavior known as mobbing. Crows and jays also mob eagles, owls, and sometimes even mammals. Their intent is to drive the potential predator from their territory, especially while they are tending their own nest or young. Sometimes their mobbing alarm calls are enough to notify a predator that it has been detected, encouraging it to move along to safer hunting grounds. If your poultry yard happens to be part of the crows' or jays' territory, in protecting themselves from predators they protect your birds as well. Each year, we lose guinea fowl eggs to crows (see Outwitting Crows [Not!] on page 235), a price we pay for the crows' diligence in patrolling our farm for hawks.

Another helpful poultry predator is the opossum. Provided you can secure your coop against 'possums — which will, if they can, snack on young poultry as well as on eggs — they will help rid the area of mice, rats, and snakes, all of which are more difficult to exclude from a coop than the larger 'possums. Mice and rats can eat an enormous amount of poultry feed, as well as spreading diseases. Rats and snakes eat both eggs and chicks. I'd rather see opossums wandering around outside my coop than find mice, rats, or snakes inside.

Coyotes, too, eat rodents and snakes, as well as foxes and feral cats. Our trail camera tells us that coyotes regularly patrol our farm, and we're happy to see them, as long as they don't bother our poultry. According to our local wildlife officer, coyotes are easy to exclude from a poultry yard, although once they discover the good things inside they become difficult to keep out. When it comes to coyotes, the old adage is true: good fences make good neighbors.

Who Eats Whom

Disrupting a predator population benefits neither you nor the predators — and, in the long run, won't benefit your poultry, either. Further, eliminating some predator species can affect populations of other poultry predators, perhaps creating a worse problem than you had to start with. Below is a list of poultry predators and the predators that, in turn, prey on them (excluding humans).

PREDATOR	ENEMIES
Bear, black	cougar, gray wolf, brown bear
Bear, grizzly or brown	none
Bobcat	cougar, coyote, wolf
Cat (feral)	coyote, eagle, fisher, fox, owl
Cougar	none
Coyote	bear, cougar, wolf
Crow	eagle, hawk, owl
Eagle	none
Fox	bobcat, coyote, eagle, lynx, cougar, wolf
Fox, gray	bobcat, cougar, coyote, golden eagle, great horned owl, wolf
Fox, red	bear, bobcat, cougar, coyote, lynx, wolf
Ground squirrel	badger
Hawk	bigger hawk, eagle, great horned owl
Magpie	crow, raven, great horned owl, hawk, weasel, mink, raccoon
Marten	coyote, fox, raptors
Mink	bobcat, coyote, wolf, great horned owl
Opossum	cat, dog, bobcat, coyote, fox, raccoon, eagle, hawk, owl

PREDATOR	ENEMIES
Otter	bobcat, coyote, raptor, alligator
Owl, barred	great horned owl
Owl, great horned	great horned owl, northern goshawk
Raccoon	cougar, coyote, dog, fox, wolf, fisher, great horned owl, hawk, eagle, snake
Rat	bobcat, coyote, eagle, fox, hawk, owl, snake, weasel
Raven	eagle, hawk, owl
Ringtail	bobcat, coyote, fox, raccoon, great horned owl
Skunk, spotted	cat, dog, bobcat, fox, coyote, owl
Skunk, striped	badger, cougar, bobcat, coyote, gray and red fox, eagle, great horned owl
Snake	bobcat, eagle, hawk, owl, weasel, fox, opossum
Weasel	coyote, fox, hawk, owl
Weasel, least	raptor, snake
Weasel, long-tailed	coyote, raptor, large snake
Weasel, short-tailed	long-tailed weasel, fox, marten, fisher, raptor
Wolf, gray	gray wolf (rarely grizzly bear)
Wolverine	bear, cougar, wolf

ELIMINATE PREDATORS
(BAD IDEA)

Despite your best efforts to protect your poultry, a particularly persistent predator — such as a fox, raccoon, or weasel — may breach your security and attack your poultry. Finding birds injured, missing, or dead is a frustrating experience that typically leads to the desire for retribution. A common inclination is to want to kill the predator, or at least send it somewhere far enough away that it can't find its way back. Since there's no end to the number of predators on the prowl, eliminating one is, at best, a temporary measure. Following are some common methods used to eliminate poultry predators, and why they are bad ideas.

Shoot, Shovel, Shut Up

A predator-control option favored by many rural folks is to stand guard and shoot. That only works if shooting is legal in your area, and if you are knowledgeable about using a firearm. A friend of mine once had a rat problem in his henhouse and thought he would be clever to go out at night and shoot the rats. When he spotted a rat running along the coop's rafters he took aim and fired, killing the rat, along with three hens that died from the concussion of the blast.

Because shooting can be messy, it has another downside: a four-legged predator could be carrying the rabies virus. Contact with saliva or brain tissue while removing the animal could result in a fatal infection, especially if infected animal fluids come into contact with broken human skin or mucous membranes of the eyes or mouth.

Before you shoot, know your local laws. If the marauder is your neighbor's dog, be sure to check local regulations regarding your obligation to notify the neighbor about your intentions. And if you're dealing with a wild animal that's protected by law, you're back to figuring out how to eliminate its point of entry.

In the long run, shooting a predator usually fails to solve the problem. For one thing, wildlife populations tend to be naturally limited by available food and water resources and suitable habitat. Depending on the species, eliminating one or more members of a given population can have the consequence that other members make up the difference by producing more young. Further, eliminating an animal vacates its ecological niche, which becomes an open

NO POISON, PLEASE

In bygone days, predators were routinely eliminated through the use of poison bait. Wisely, this practice has been mostly discontinued because it too often results in the death of nonpredatory animals, some of which may be valuable livestock or family pets.

invitation to another member of its species to move in. Finally, predators that get shot tend to be the least wily, leaving the cleverer ones to live and reproduce more of their kind. In all cases, eliminating a predator could cause the predation problem to become worse that it was to start with.

Trap and Move

If all your efforts to resolve a poultry predator problem prove futile, a final alternative is to trap the predator. It is the final alternative because to trap legally you will likely need a permit, which will be issued only if you can demonstrate that all nonlethal methods of predator control have failed.

Wildlife traps are of three types: quick kill (such as a snap trap for mice or rats), restraining (such as a leg snare or leghold trap), and cage-style live traps. The latter are the type most often used for dealing with poultry predators, and in some areas they are the only kind of trap allowed by law, to avoid injuring nontargeted species, including people.

A popular belief is that the humane way to deal with a predator is to capture it in a live trap and remove it to a distant location. This solution is a bad idea for the following several reasons.

Trapping can be dangerous. A trapped animal may try to bite you when you move the trap or may spray in defense. It may be rabid or otherwise ill, contaminating the trap and surrounding area with dangerous disease-causing organisms. A sick trapped animal that is released elsewhere will spread disease to otherwise healthy populations.

Trapping is subject to error. You may catch the wrong animal — maybe not the animal that has been attacking your poultry, or maybe even your own pet. Trapped wildlife sometimes escape from traps, either out of cleverness or because the trap didn't function properly. In such a case, the animal becomes trap smart and will be more difficult to catch in the future.

What to do with the trapped animal? Where, and how far away, can you safely and legally release it? In some cases, the law requires a trapped animal to be either released on the same property where it was trapped or euthanized. If the latter is feasible and legal, you must have a plan for disposing of the dead animal. If you can't, or don't want to, do it yourself, a local wildlife damage control company, veterinarian, or animal shelter may euthanize the animal for a fee, although many veterinarians and animal shelters will not handle sick wildlife for fear of spreading disease.

Trap-and-move is cruel. A relocated animal won't know where to find food, water, and safe shelter and therefore likely won't survive for long. Further, most predators have established territories. An animal moved outside its territory will do everything in its power to return to its home territory. In doing so, it may pass through territories established by other members of its species, and they will attempt to fight off the intruder. Animals will be injured or killed. An injured animal may not be able to defend itself in future fights and may not have the ability to hunt or forage, and therefore will starve. If the trapped predator was pillaging poultry to feed its young, the babies will either starve or themselves fall victim to some other predator.

Trapping doesn't solve the problem. Should a released predator make it back to its home territory, most likely it will revisit your poultry yard — but now it is trap smart and will be hard to catch again. Even if the original predator doesn't return, some other member of its species will eventually move into the vacant territory. If your poultry yard remains unsecured against predation, the problem will recur and may even get worse.

Trapping is not easy. Cage traps come in many styles and sizes for trapping different species. Unless you know what species you are targeting, already have access to a trap, and are experienced using it, trapping can be difficult and time-consuming. An experienced trapper spends years becoming familiar with predator behavior patterns, how to entice an animal into a trap, how to recognize signs of disease, and the legalities involved in predator trapping and disposal. If you feel trapping is necessary but you don't feel qualified, you might locate a wildlife damage control company.

Trapping may be illegal. Species that are common in one state may be listed as endangered in another state. In many states, possessing or transporting live wildlife without a permit is unlawful. Some species are protected by law from being trapped or killed at all. Before you can trap and kill or relocate a predator, the law says you must first contact your local wildlife office and municipality for current information on trapping restrictions (the type of trap you may legally use,

▲ Eliminating a predator may leave its baby without a protective parent.

euthanization requirements, biological concerns) and to obtain any required permits. Illegal trapping can result in prison time and a hefty fine.

Legal trapping is generally permitted only after you have exhausted all other possibilities for avoiding significant poultry predation. In such a case, you may qualify for a depredation permit, the intent of which is to provide short-term relief while you implement long-term nonlethal measures to eliminate or significantly reduce the problem. The easiest and most effective solutions to a poultry predator problem are to secure your birds by eliminating points of entry and to remove attractants to discourage the presence of wildlife. If these things are accomplished at the outset, you won't have a predator problem.

See the Resources for a list of state and federal agencies that can help you deal with predators.

5. FENCE DEFENSES

"It is simpler to deal direct."

Sherlock Holmes

Short of keeping your poultry flock locked inside their coop 24/7, which isn't at all healthful for the birds, your best defense against predators is a securely fenced yard. Even if your flock is small enough to be housed in a prefab coop that comes complete with its own fenced run, you may want to add a perimeter fence as additional security against harassment by dogs, coyotes, raccoons, and the like and, if any livestock are present, having the coop rubbed on, climbed on, and bumped into. The larger the poultry yard, however, the more difficult it becomes to make it secure against predators.

FENCE SELECTION

Regardless of its size, a secure fence can be expensive to construct, so it pays to do plenty of advance research and planning. The first place to start is with your city or county planning commission. Since local zoning laws may restrict your choice of fences, check for regulations in your area that pertain to fence design and construction. If you put up a new fence that doesn't conform to local restrictions, you may end up having to take it down and possibly also pay a fine.

Where an agricultural-style fence or electrified wires are not permitted, a solid wooden fence, a high picket fence, or an attractive rail fence backed with less visible wire mesh are some alternatives. Even where no local regulations restrict your choice, choose a style of fence that blends in with its surroundings. Also consider ease of maintenance — spending more time and money now may save you lots more time and money spent later on repairs.

Selecting the ideal fence isn't always easy, since each system has inherent strengths and weaknesses. And poultry offer unique challenges because they can be small enough to slip through a fence or light enough to fly over it. More difficult than keeping a flock in can be keeping predators out. Just how persistent predators will be often comes down to how hard they have to compete for food and water outside the fenced area, which keeps changing with such things as predator population dynamics and weather conditions. So construct your fence for the worst-case scenario.

Gates are often overlooked as potential security breaches — if chickens can get out, predators can get in.

MESH FENCES

One of the best fences for poultry is constructed of tightly strung woven wire with a closely spaced mesh. Woven wire is sometimes mistakenly called welded wire, which is an entirely different thing. Welded wire mesh consists of horizontal and vertical wires welded together wherever two wires meet. The result is a rigid, inflexible mesh. Welded wire is commonly used to make cages, such as those used to transport poultry or to house birds at a poultry show.

Other examples of welded wire are hardware cloth and livestock panels. Hardware cloth is suitable for making a chicken tractor or a fully enclosed poultry run that, in turn, is enclosed within a perimeter fence providing a free-range area outside the tractor or run. Livestock panels are much sturdier than hardware cloth and may be used to create a fast and secure fence, as described on page 111.

Welded fence wire that comes in rolls, however, is suitable only on perfectly flat terrain because of its inflexibility. Additionally, if a fence stretcher is used improperly to tighten the mesh between fence posts, welds may pop. Welds can also pop under pressure from livestock or a persistent predator pushing against the fence.

Like welded mesh, woven mesh is made from horizontal and vertical wires, but the wires are woven, knotted, or hinged together instead of being tightly welded where they meet. The resulting flexibility allows the wire to readily adjust to uneven ground and to handily accommodate a fence stretcher. Examples of woven mesh are chicken wire, chain link, and field fence.

Chain Link

Right at the beginning of my poultry-keeping career I got spoiled, because my first house came with chickens and ducks in a yard fenced and cross-fenced with

FORGET CHICKEN WIRE

A type of woven wire mesh fence that is anything but predator proof is variously known as hexagonal netting, hex net, hex wire, or poultry netting. Depending on the mesh size, it's also called turkey wire, chicken wire, or aviary netting. It consists of thin wire, twisted and woven together into a series of hexagons, giving it the appearance of a honeycomb. All-plastic versions are also available. Both the wire and the plastic mesh make a lightweight fence that has a limited life span and does not deter predators, many of which have no trouble chewing through it or tearing it apart. The best use for this type of netting is inside a protected coop to create individual areas for separating birds by age, species, or whatever your criteria might be. Bottom line: hexagonal netting will keep chickens in, not predators out.

6-foot-high chain link. In the 11 years I lived there, I lost few birds, most of which were babies that got into harm's way by popping through the fence to go exploring independent of their protective parents.

I now live on a farm where we use a variety of fencing options, my favorite of which remains chain link. Two of our poultry yards are protected by 7-foot-high chain link fences. One of the fences surrounds an outdoor brooder where we raise chicks in warm weather. The other is variously used for protecting bantams or replacement layer pullets.

To prevent birds from sneaking out and predators from pushing underneath, the bottom of our chain-link fence is fastened to 8-inch-wide pressure-treated boards turned on edge. Along the top of the fence we ran a wire strand to which we attach DVDs, as described on page 81–82. Outside the fence, to keep predators from climbing

on it, is a single scare wire, as described beginning on page 112.

Standard chain link mesh size is 2 inches. So-called mini mesh comes in several smaller mesh sizes, the smallest of which is ⅜ inch. Although mini mesh is more expensive than standard mesh, it is also much more difficult to penetrate or climb and therefore is ideal for building predator-proof and escape-proof pens to protect baby poultry.

A roll of chain link contains either 25 or 50 feet. Heights range upward from 36 inches. Standard chain link is woven from 11-gauge steel wire and may be galvanized, aluminized, or vinyl coated in a selection of attractive colors.

The mesh is fastened to steel posts, set in concrete and evenly spaced 8 feet apart, or 10 feet if a steel top rail is used to keep the mesh taut. Amateur installers often omit the rail because it requires

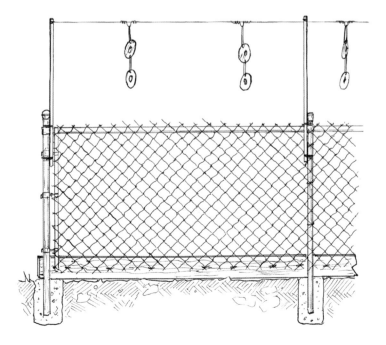

◄ To enhance the protective quality of a chain-link fence, fasten the bottom to a rigid board and hang wind catchers along the top.

the line posts to be all set at exactly the same height and perfectly plumb. If you choose to leave off the rail, top off all your line posts with caps so they won't collect rain water.

Consider installing your own chain-link fence only if you have a knack for meticulous detail — this fence has zillions of little parts to fit together. Look for a store that encourages owner-installed chain link, where you can get all the tools and moral support you need, as well as learn the lingo. In chain link parlance, for example, an end post is called a "terminal" post and chain link mesh is called "fabric." When you buy materials, the supplier will help you determine the size, kind, and number of parts you need. Bring along a layout showing exact dimensions, locations of corners and gates, and fence height.

Given the cost of materials, you might be happier with the results if you hire an experienced installer, especially if your yard is anything but level with right-angle corners. A professional fence installer has all the necessary equipment, plus the knowledge to get the job done correctly and quickly. To save a few dollars on the cost, you might ask to be hired as a helper.

Although chain link is an expensive option, when correctly constructed it is virtually maintenance free and excludes most predators. Its chief disadvantage is that you'd better know exactly where you want your fence before it goes up. Deciding later to move a fence or expand the yard can be considerably costly.

Field Fence

If the cost of chain link is more than you care to absorb, the next best alternative is a small-mesh field fence, also known as yard fencing, lawn and garden fencing, small stock fencing, goat and sheep fencing, kennel fencing, and sometimes even poultry fencing. Typical mesh for poultry yards is either 2 inches square or 2 inches wide by 4 inches high.

To discourage rust, field fence is galvanized with a zinc coating. Standard galvanizing, called class 1, lasts about 10 years before rust sets in. Class 2 galvanizing is two and a half times thicker and typically lasts three times longer than class 1.

Most brands of field fence are made of medium hardened steel. The hardness of the steel affects the amount of springiness built into the fence. Soft steel has little intrinsic springiness and will sag no matter how carefully you stretch it.

Spring action is developed through tension curves — wavy crimps that stretch when the fence is pulled taut. Tension curves give the horizontal line wires room to stretch and shrink as the weather changes. Without them, properly stretched wire would either break or pull out anchor posts in cold weather.

Woven wire rolls generally come in 3-, 4-, 5-, and 6-foot heights. For most poultry the fence should be at least 4 feet high. Make it higher if you keep a lightweight breed that likes to fly. Make it higher also if your problem predators are bobcats or coyotes, both of which can easily leap over a 4-foot fence. And if you plan to cover the top of the run, as described on page 75–77, do yourself a favor and make the fence at least 6 feet high to avoid having to crouch whenever you work in the run.

The taller the wire, the more difficult it is to work with. If you prefer to install your own fence, you might opt for the 4-foot height, then increase the height by stringing one or more scare wires along the top, as described beginning on page 112. If you prefer the taller mesh or you have lots of fence to build, you might be happier to hire an experienced crew with all the necessary equipment to get the job done fast and right. A good place to start your search is a farm store or builder's supply outlet where field fence wire is sold.

Attach the wire mesh to the side of the posts experiencing the greatest amount of routine pressure, putting the brunt of pressure against the posts rather than the fasteners. Where chickens and livestock share the same yard, fastening the wire to the inside of the fence posts will help protect the fence from the larger stock. Where the sole purpose is to keep predators out of a poultry yard that does not include larger livestock, fasten the wire to the outside of the fence posts.

Plastic Mesh Fence

Plastic mesh fencing is less expensive than metal, doesn't rust, is lighter and therefore easier to handle, may be cut to size with scissors, and may be attached to posts with zip ties. Its chief uses are as a yard barrier to keep poultry out of certain areas, such as a vegetable garden, flower bed, or patio; to contain baby chicks within a larger fenced area; to provide a temporary, movable run; to cover a run as protection from predatory birds; and as apron fencing to discourage diggers.

The mesh may be woven from plastic twine or may consist of a solid sheet with patterned holes. The latter is sometimes oriented, or stretched, in two directions for increased strength. Mesh that's been oriented in only one direction is cheaper, but not as strong. Mesh that hasn't been oriented at all is cheapest and weakest. If you construct a plastic fence in summer, allow slight slack to accommodate cold-weather contraction. In winter, stretch it as taut as possible, since the plastic will expand come summer.

One disadvantage to plastic mesh fence is that it is subject to ultraviolet deterioration. To resist UV deterioration, the manufacturer may add a UV inhibitor. The more inhibitor the plastic contains, the more it costs and the longer it will last, ranging from as little as 3 years to 10 years or more. Of the various available colors, black is the most resistant to ultraviolet deterioration, a factor that becomes increasingly more important the farther south you are.

The main disadvantage to plastic mesh as a protective fence is that it is easy for predators — and poultry — to rip it apart or chew through it. For keeping out predators, an electrified version of plastic netting is available, as described beginning on page 110.

DEFEATING DIGGERS

Keeping predators from digging underneath a poultry fence can be a challenge. If your birds stay out all night, they certainly will attract dogs, coyotes, foxes, raccoons, and other diggers. Even if you lock up your birds at night, diggers may still try to get in. And once a gap has been created underneath the fence, your birds will use it as a handy way to slip out.

How much effort and expense you wish to put into discouraging diggers depends a good deal on how long you expect your fence to last. If you are planning what you expect to be a lifetime fence, you might want to go all out. Here is a range of possibilities:

▶ **Build your fence on top of a deep concrete footer.** That was the solution my family chose when we built the first coop on our farm. It was time-consuming and expensive, but it worked. It was also permanent. A little too permanent, as things turned out. When we decided to move the chicken coop, the concrete footer was problematic until we devised a way to incorporate it into our raised-bed garden.

▶ **Lay paving stones.** A more portable option is to lay large flat stones or concrete patio pavers on the soil's surface along the outside of the fence or wall. Diggers are then forced to start digging farther away and may give up before they get through.

▶ **Bury the bottom 12 inches or so of your fence.** This solution, like the first one, requires digging an extensive trench. It also requires using taller fencing, because part of it will be underground. Further, the fencing material must be super rust resistant, or you will need to coat the bottom portion of the fence with a rust-resistant paint or a compound such as roofing tar.

▶ **Install Dig Defence panels.** This option is not cheap, but the panels work well and they're easy to install — provided your soil, unlike ours, is not full of rocks. Each panel consists of a series of parallel steel spikes that you push into the ground by means of a top rail to which the tines are welded. Dig Defence panels (see Resources) come in several styles ranging from 8 inches deep with spikes 2 inches apart for deterring smaller diggers to spikes 15 inches deep and 1½ inches apart, which will deter nearly any digger.

▶ **Add an apron.** This option is also known as a skirt fence, beagle net, or L footer. It consists of a wire mesh barrier at least 12 inches wide, securely attached at a right angle along the outside bottom of the fence. Since diggers normally start at the base of a fence, they have trouble getting through or underneath the apron. Commonly recommended for apron fencing is hexagonal netting, which does not hold up nearly as well as a sturdier mesh such as hardware cloth. Mesh openings of 2 to 3 inches will keep out most diggers except rodents and weasels, for which you will need openings of 1 inch or less. The apron needs a good rust-resistant coating, since even galvanized wire begins to rust almost immediately upon contact with moist soil. Although typically the apron is buried under a foot of soil, it may be laid on bare soil and covered with sod, or even spread right on top of the ground. In the latter case, to prevent raccoons and other clever wildlife from lifting the apron and crawling underneath, and to facilitate mowing along the outside of the fence, securely anchor the apron with landscape staples, available where garden supplies are sold.

► **String an electrified scare wire.** This option is described beginning on page 112.

► **Create a pee-rimeter with predator urine.** According to PeeMan Ken Johnson (see Resources), the poultry predator you want to discourage will be deterred by the urine scent of the predator it, in turn, is most afraid of (see Who Eats Whom on page 86). Accordingly, to deter bobcats, coyotes, and foxes use wolf urine. To deter opossums, raccoons, and rats use coyote urine. For dogs use bear urine. For skunks use fox urine. Place the urine about every 10 feet — which, conveniently, is about the distance between fence posts — and reapply at least once a month or after a rain. This solution can become time-consuming and costly and therefore might be considered as a temporary measure.

apron fence

concrete footer

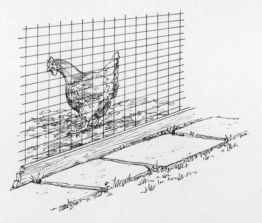

paving stones

DEFEATING JUMPERS AND CLIMBERS

Some predators can handily climb or jump over a fence. A coyote, for instance, can easily jump an 8-foot fence. Where the law does not allow electrified scare wires (see page 112), you still have a couple of options.

One is to use fence extension arms to expand the top of the fence upward and outward. When a climbing or jumping animal encounters extension arms, gravity causes it to drop back to the ground.

Another option is to install roller bars along the top of the fence. When an animal attempts to scale the fence, the roller bar will turn in the direction the animal is coming from, causing the animal to lose its balance and fall back. Ready-made roller kits such as Roll Guard (see Resources) are available, or you easily make your own following any of several online roller bar DIY instruction sites.

To deter high-jumping coyotes or bobcats, the fence should be at least 6 feet high to ensure that the animal encounters the extension arms or reaches for the roller bar, instead of sailing right over the top of the fence.

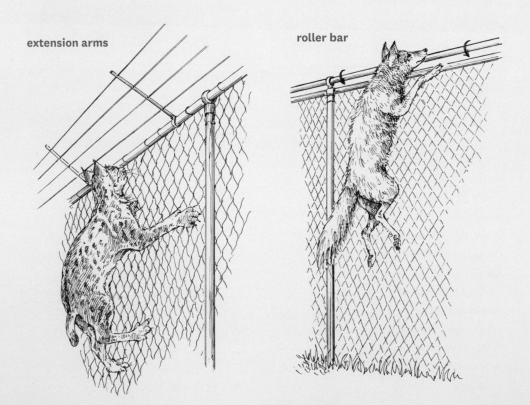

extension arms

roller bar

ELECTRIC FENCES

Compared to most other fences, an electric fence is easier and cheaper to build. It also lasts longer, requires less maintenance, and is less easily damaged than most other fences, since animals and people don't climb or lean on it. A properly constructed electric fence also offers greater predator protection than nearly any other kind of fence. Unlike a mesh fence, which is a physical barrier through which, theoretically, poultry can't get out and predators can't get in, an electric fence is a psychological barrier — it sufficiently intimidates predators into not wanting to get in, without causing them physical harm.

An electric fence does, however, have certain disadvantages. Unless it's well enough constructed to function as a physical barrier, it must be electrified at all times to remain effective against predators. You'll also need to keep the fence free of shorts (short circuits, as described on page 109), which reduce the fence's wallop.

Another disadvantage is that the fence's electrical conductivity can vary with climatic conditions. Design your fence with this factor in mind, or the fence will become less effective in dry or snowy weather and could easily short out in wet weather.

Finally, in some areas the law prohibits electric fences, either at all or in certain circumstances, such as adjacent to roadways or public lands. Check your local ordinances before constructing an electric fence on your property.

To be predator proof, electrified fence wires must be high enough that animals can't jump over, spaced closely enough that they can't push through, and low enough that they can't dig or crawl under. To deter animals with thick fur, the fence must also pack one heck of a wallop.

Wire designed for building an electric fence is made of either steel or aluminum. Aluminum is more conductive than steel. It is also more flexible, thus easier to work with, but will break if bent repeatedly. Steel wire is stronger and virtually indestructible, and it is typically used for high-tensile fences (see page 108 for a discussion of tensile strength). In both cases, wire for fencing should be at least 14 gauge; the smaller the gauge, the thicker the wire.

Selecting an Energizer

An electric fence is powered by means of a unit called an energizer. The most common type of energizer is a plug-in (or mains) energizer, which plugs directly into a standard 120-volt household (AC) electric outlet. Where electricity is not available, or power outages are frequent, battery-operated (DC) energizers are available, some of which have built-in solar panels. A plug-in unit is generally more reliable and, because it consumes low wattage, is relatively inexpensive to operate.

The purpose of the energizer is to convert a steady electrical output into short bursts of energy pulses. It does so by reducing the electrical rate of flow (amperage) while increasing the electrical force (voltage) that drives the electric current along the fence wires. High amperage is what makes electricity deadly dangerous. Voltage provides the shock you feel if you are unlucky enough to touch an electric fence. Although the jolt is memorably uncomfortable, it does not cause damage or injury because of the low amperage.

An energizer releases a pulse about once every second, with each pulse lasting less than 0.003 second. Although the high voltage results in a painful a shock, the off time between pulses gives you (or a predator) a chance to pull away from the fence.

Selecting a suitable energizer for your particular fence can be confusing. For one thing, you need to be familiar with the terminology commonly used in describing an energizer's output. Further, not all manufacturers measure the output of their energizers in the same way. If the

▲ High-energy low-impedance electric fence energizer

technical details make your head hurt, contact a knowledgeable dealer of electric fence supplies, who can recommend the best energizer for your purpose based on a detailed description of your fence. We find that Kencove Farm Fence Supplies (www.kencove.com) both maintains an informative website and offers outstanding customer service.

Volts and Joules

The output of some energizers is indicated using the misleading term "miles of fence." This phrase refers not to the length of your fence but to the total length of all the line wires making up the fence. Further, this number is derived by measuring output in a single continuous strand of wire 3 feet off the ground and free of weeds — not exactly a real-life scenario. The actual distance a pulse can travel in field conditions is considerably less because of the influences of climate, weed load, fence construction, wire gauge, and the electromagnetic resistance that arises when electrified wires are strung parallel to each other.

The output of most energizers is rated in volts and joules. Volts measure the amount of force behind each pulse. The more voltage an energizer puts out, the more readily it leaks energy through such things as weeds and faulty insulators. On the other hand, the higher the voltage, the farther a spark will fly from the pulsed wire to a predator's coat, and the greater the chance it will penetrate that coat. To impress most predators requires between 5,000 and 10,000 volts. High voltage is necessary for stopping furry animals like coyotes and bears. However, even though a predator with a thick coat might bump against a 5,000-volt wire and feel nothing, if the same animal touches the wire with its nose, tongue, or ear, it likely won't come back.

Joules are units of energy, an important consideration when buying an energizer. One joule is the amount of energy needed to push 1 amp of current through a 1-volt force for one second. When looking at an energizer's joule rating, take note of whether the figure is joules stored or joules delivered. Stored joules tell you the amount of energy stored in the energizer's capacitors. The important number is output joules, which tells you how much energy is actually released from the energizer to power your electric fence. If the only information you have is stored joules, figure the output will be about 70 percent of that.

As a general guideline, an energizer that deters predators requires at least 2.8 output joules per mile of line wire. Factor in extra power if, like most of us, you don't diligently keep your fence weed free. And if you use polywire you will need more joules than if you use metal wire. My recommendation is to get the next-highest joule energizer than you think you need.

On our farm we have two similar areas protected by electric fence, both of which are shared by poultry and dairy goats. One area has a perimeter fence powered by a 9-joule energizer. The other has a 13-joule energizer. Of the two, the 9-joule energizer is much more sensitive to impedance, meaning the 13-joule energizer does the better job.

ELECTRIFYING WORDS

Arcing. Sparks indicating that pulses of current are jumping across a gap.

Current. The rate of flow of an electrical charge through wire, measured in amperes (amps).

Dead short. A direct connection between energized line wires and the ground or grounded wires.

Hot wire. An energized line wire carrying live current.

Impedance. The measure of how much resistance is present to restrict energy flow.

Joule. A unit of energy (power) representing 1 amp of current through a 1-volt force for one second.

Line wires. Strands of horizontal wire strung parallel to each other along the length of a fence line.

Polywire. Also called electroplastic twine. Fence material consisting of several polyethylene strands twisted together with stainless-steel filaments for conductivity.

Short. A short circuit, or voltage leak.

Stray voltage. Electricity in places where it shouldn't be.

Voltage. The force that drives an electric current through wire.

Energizer Impedance

Pulsed current zips effectively along a fence wire unless it encounters resistance that restricts its energy flow. The measure of how much resistance is present is called impedance. The greater the resistance, the higher the impedance, causing each pulse to gradually dissipate as it moves along the wire, reducing voltage and rendering the fence less effective. Or perhaps completely ineffective, if the voltage falls so low that it can't arc across a predator's insulating fur coat.

Impedance relates not only to resistance but also to the amount of current that flows with a specific voltage. A low-impedance energizer pushes more power through the fence wires (compared to now-obsolete high-impedance energizers) and therefore is better able to shock through a modest amount of vegetation and debris along the fence. Still, the unit works best when resistance is low, which occurs when the weed load is light, the soil is moist, and plenty of grounding rods (see page 109) have been installed.

The best source for a low-impedance energizer is a firm with friendly customer service and a reliable repair department. Look for an energizer with replaceable parts, because it will be much less expensive to repair than to replace. And make no mistake, you can expect your energizer to require occasional repair, especially if your area experiences frequent lightning storms.

Because breakdowns are inevitable, a good warranty is a big plus. Some manufacturers warrant their low-impedance energizers for as long as three years, and some even include damage to the energizer caused by lightning.

Unless it's fried by lightning or disabled by a power surge, a good low-impedance energizer should last at least 10 years but will likely need repair every 2 to 5 years. You can prolong the life of your energizer by properly grounding your system, by using sound insulators and keeping them clean, by keeping your fence cleared of weeds, and by installing lightning diverters (see the facing page).

Energizer Accessories

Depending on the style and length of your electric fence, and whether your area is subject to frequent lightning storms, some of the following accessories may help simplify maintenance of your energizer:

Fence tester. A handheld battery-operated fence tester can tell you if your electric fence is working right and help you find shorts if it isn't. An inexpensive tester is simply a voltmeter. A high-end tester also displays amperage.

Remote fence tester. This type of tester not only displays voltage and amperage but lets you remotely turn off the energizer while you make fence repairs. A remote may be used with only the specific type of energizer for which it has been programmed.

Dead fence alert. A battery-operated fence monitor, or fence alert, clips to a hot wire and flashes a light if voltage drops on the fence or the energizer stops working, letting you know that the fence is no longer effective. It not only notifies you of the problem but, as a

Fence tester

Dead fence alert

flashing red light, helps ward off predators until the problem is resolved.

Live fence indicator. The function of this device is opposite that of the dead fence alert. Instead of notifying you if the voltage is low, it flashes when the fence is functioning properly. Unlike the fence monitor, which requires a battery, this unit uses energy from the fence itself. The main decision here is whether you want something that flashes all the time or only when a problem arises.

Surge suppressor. If your plug-in energizer doesn't have built-in surge protection (most don't), you can minimize repairs by protecting it with a surge suppressor, available from your energizer supplier. Keep a spare on hand, as the suppressor will need to be replaced whenever it gets damaged by a power surge.

Lightning diverter. A lightning diverter protects your energizer from a lightning strike traveling along your fence by grounding the lightning before it reaches the energizer. When a diverter made of porcelain does its job, it cracks

(after which you will hear it click with each pulse traveling along the fence) and must be replaced.

Cut-off switch. When placed in the fence near your energizer, this switch lets you disconnect the fence to protect your energizer from lightning strikes. When placed at a distance from your energizer, it lets you disconnect the current to a portion of your fence while you make repairs. When placed on the fence's lowest wire, it lets you disconnect the wire if it comes under a heavy weed load or gets covered in snow, in which case disconnecting the bottom wire improves voltage on the rest of the fence.

Warning signs. Bright yellow signs printed with the words "Electric Fence," sold by farm stores and suppliers of fencing materials, are required by law in some areas. Your county Extension agent should be able to tell you where and how signs must be posted to comply with local laws. Posting signs, even where the law doesn't require it, is a good idea for safety's sake.

How It Works

For an electric fence to work properly, current must flow from the energizer through the line wires and back to the energizer through the soil, closing the circuit. Under normal circumstances, electrically charged wires are separated, or insulated, from the earth and the circuit is open. A predator standing on the ground and coming into contact with a pulsed wire closes the circuit and feels the consequences.

A chicken or guinea fowl slipping through the fence is unlikely to get a shock because its feathers insulate it from the hot wire and its feet don't provide sufficient contact with the ground. A songbird landing on a hot wire doesn't feel anything because it doesn't complete the circuit. On the other hand, if a little bird sitting on a wire pecks a bug off a metal fence post, it's bye, bye birdie. One of the hazards of electric fencing is that lizards, praying mantids, and other small creatures get fried when they crawl up a metal post and touch a hot wire, thus closing the circuit. The same jolt that's lethal for creatures of small mass, though, is merely unpleasant for a larger animal.

When the circuit must be completed by an animal standing on the ground, the system is called a ground-return, earth-return, or all-hot system. An earth-return system works fine for relatively short runs of fence in areas of even rainfall where the soil is readily conductive and predators are not a persistent problem. It does not work well where the soil lacks conductivity because it is dry, sandy, rocky, frozen, or insulated with a layer of packed snow.

Further, a predator that gets through a fence by leaping between the wires doesn't feel a shock because its feet are off the ground.

A wire-return, or hot/ground, system works independently of weather and soil conditions and is effective even for long fences. It deters animals with thick fur and those performing tricky maneuvers like jumping through a fence instead of crawling under it or attempting to climb over it.

In a standard wire-return system, every other wire is connected to ground. Typically, the second and fourth wire from the bottom are electrified to discourage digging and crawling, the top wire is electrified to discourage climbing, and the spacing is adjusted so the wire closest to nose height of the animal is electrified to discourage pushing or looking through the fence. When an animal touches two adjacent wires, one hot and one grounded, it completes the circuit and feels a jolt. Even if the animal hits the fence with all fours off the ground, it needs only to touch two wires to close the circuit.

Exactly which wires should be grounded is a matter of debate. Some fences have the bottom wire grounded to reduce energy leakage from encroaching weeds. Others ground the second wire up as a better deterrent to young stock and small dogs. With the bottom wire hot, an animal pushing under the fence has two ways to complete the circuit — by touching the earth and the bottom wire or by touching the two lowest wires.

Also debated is whether or not to ground the top wire. On the open plains, grounding the top wire reduces the risk of lightning damage to the energizer,

Earth-Return versus Wire-Return Electric Fence

EARTH RETURN (ALL HOT)
- Effective if predator touches one wire
- Requires good soil conductivity
- Needs fewer line wires
- Less susceptible to shorts
- Easier to maintain
- Effective over short distances

WIRE RETURN (HOT/GROUND)
- Effective in multiple ways
- Independent of soil conductivity
- Needs more line wires
- More susceptible to shorts
- More time-consuming to maintain
- Effective regardless of fence length

◄ Earth-return system, also known as all-hot

► Wire-return system, also known as hot/ground

because it diverts a strike directly to earth with little damage done. On the other hand, where trees are present, lightning typically strikes a tree and then arcs to the fence, thus hitting more than just the top wire. Whichever wires you choose to ground, always follow these important rules to ensure the effectiveness of your wire return system:

- Energize the line wire nearest the nose height of the predator you intend to deter.

- Connect the hot system and the ground system separate from each other.

- String the wires at least 4 inches (10 cm) apart so they can't inadvertently touch.

An electric current always produces a magnetic field. The stronger the current, the more intense is the magnetic field. Spacing wires too close together could result in stray voltage, which is caused by induction — the process by which electricity from an electrified wire is transferred to a nonelectrified object (such as a ground wire or metal gate) without having physical contact. I have had unpleasant encounters with stray voltage, once when I tried to open a metal gate that wasn't supposed to be hot, and another time when I simply sat down on the ground. (You can guess what that was like.)

For predator control, attach insulators on the outside of the fence posts. Insulators should firmly hold line wires to preserve their distance from each other and from the post, yet allow the wires to move freely in case of sudden impact.

Line Wire Spacing

The spacing between line wires in an electric fence is indicated by a series of numbers separated with hyphens. A five-wire fence might have this designation: 12-10-10-8-8 (in metric: 30-25-25-20-20). The first number tells you, in inches or centimeters, how high the bottom wire is from the ground. The second number tells you the distance between the first and second wire, and so forth. To determine how far the top wire is from the ground, add up all the numbers. To calculate the length of the posts you'll need for this fence, add together the height of the fence plus 6 inches plus the depth at which the posts will be set, typically 3 to 4 feet.

No firm and fast rules have been established for exactly how many line wires a fence needs and how far apart they should be spaced to deter specific types of predator. The possible combinations of climate, terrain, livestock type and temperament, and predation problems are virtually endless. The same fence that works perfectly for one person may require minor or major adjustments for a seemingly identical situation elsewhere.

Happily, experienced electric fence users have developed a few good guidelines. For starters, the most important hot wire is the one at normal nose height for the animal being controlled, or about two-thirds the animal's total height. The bigger the predator to be controlled, the taller the fence must be and the more line wires it needs.

Even though wire is the cheapest part of an electric fence, stringing too few wires is a common error. The addition of just one more wire could double a fence's effectiveness while adding minimally to its cost. Provided the wires remain at least 4 inches apart, the more line wires your fence has, the better it will work as a physical barrier as well as a psychological barrier. When in doubt, add more strands than you think you need and you won't end up with too few.

In general terms, a six-strand wire-return system will keep out marauding foxes, as well as repelling casual coyote and dog incursions. Where dogs and coyotes are a significant threat, you'll do better with an eight- or nine-strand wire-return system having a total height of about 45 inches. Since canines search out gaps, keep the bottom wire tight, hot, and following the land's contour no more than 6 inches above the ground. A 10-strand wire-return tension fence (see the next page) will keep out most predators and double as an effective physical barrier should the power go out. Since the bottom one-third of any fence experiences the most animal pressure, line wires are typically spaced closer together toward the bottom and progressively farther apart as the fence reaches optimum height.

The following table indicates line wire spacings that most commonly work for the predators indicated. If your fence uses wood or fiberglass posts, you can position insulators wherever you wish and follow the specifications exactly. If you use steel T-posts, you'll need to adjust wire spacing so the insulators will fit between the post's lugs.

Electric Fence Line Wire Spacing

PREDATOR	HEIGHT	WIRES	SPACING IN INCHES	SPACING IN CM
Raccoons + small animals	13"/33 cm	3	4-4-**5**	10-10-**13**
Bears	40"/100 cm	5	4-5-6-15-**10**	10-12-15-38-**25**
Fox	39"/98 cm	6	2-6-6-8-8-**9**	5-15-15-20-20-**23**
Coyote, bobcat, cougar, fox	46"/117 cm	8	4-5-5-5-6-**6**-7-**8**	10-13-13-13-15-**15**-18-**20**
Most predators	47"/120 cm	10	4-4-4-4-5-**5**-5-**5**-5-**6**	10-10-10-10-13-**13**-13-**13**-13-**15**

*Wires in **bold black** are energized; others are grounded.
Adapted from *Fences for Pasture and Garden* by Gail Damerow (Storey Publishing, 1992)

Tensile Strength

The tensile strength of metal fence wire measures how much it can be stretched before it breaks and is designated as either high or low. High-tensile wire has built-in springiness and can be mechanically stretched much tighter than low-tensile wire. Low-tensile wire, often called soft wire because it contains less carbon and therefore is physically softer, can be stretched only enough to eliminate sag.

High-tensile wire must be used to construct a high-tension fence, commonly referred to as a tension fence. The word *tension* refers to the wire's tautness. A tension fence is strung taut enough that the wires will spring back into their original position after a sudden impact, and a predator can't easily pry the wires apart to slip between them. If a predator tries to force its way between tensioned wires on an electric fence, it will be discouraged by shocks to the nose and ears, causing the animal to pull back. If the predator is partway through a low-tension fence before it feels a shock elsewhere on its body, it will jump forward and get through.

A low-tension electric fence is easier and cheaper to construct because it doesn't require the multiple supports and tensioners needed to produce high tension, but it's not nearly as effective because the wires can be more easily spread apart. Further, soft wire, instead of bouncing back, stretches and breaks on impact. A low-tension electric fence may be constructed of polywire, which is cheaper and easier to string than metal wire but breaks or wears through more easily and therefore won't last nearly as long. It is also more resistant and therefore requires more energy per mile to effectively electrify.

A tension fence isn't for everyone. Installing one on hilly terrain or where the fence line consists of short runs with lots of corners, for example, would be extremely costly because it requires considerably more fence posts than are needed on level terrain with long, straight runs of fence. And you certainly wouldn't put one up if you didn't intend the fence to be permanent — constructing a tension fence entails a lot of work, and removing one isn't much easier. If you're at all handy you could likely install a low-tension fence yourself; a high-tension fence should be constructed by someone knowledgeable about what it takes to keep the fence adequately tensioned.

Two of our poultry flocks are protected by high-tension wire-return electric fences, because in both cases the yards are shared with dairy goats. The first fence was constructed in 1990 and is still as strong as the day it was built. Like our chain-link fences, our tension fences require little maintenance, other than controlling weed growth along the fence line, and they do a terrific job of keeping out predators. Unlike chain link, however, the tension fences don't prevent our chickens from slipping outside their yard. For us that's generally not been a problem, because the chickens just cross the driveway to forage in our orchard; when they feel threatened they scurry back to the safety of their fenced yard.

Grounding Rods

The success of any electric fence depends on proper grounding, which literally means providing the current an easy path through the soil. Far too many electric fences have inadequate grounding systems, resulting in poor predator control, not to mention voltage leaks, blown fuses, and damaged energizers. Good soil contact can be provided by means of ⅝-inch-diameter grounding (sometimes called earthing rods) made of carbon steel that's galvanized or copper plated to resist corrosion.

Grounding rods must be driven deep enough to make good contact with damp earth during your driest season. The ideal place for a rod is a shady low spot that's moist year-round, such as along a building's drip line or north corner where

SHORT CIRCUITS

Equally as important as establishing an adequate grounding system is making sure the hot wires have no contact with the ground or grounding system. A direct connection results in what's called a dead short, which occurs when a hot wire touches the earth, a grounded wire, or a metal fence post.

Lesser contacts, called shorts or leaks, occur when a hot wire touches something less conductive, draining energy from your fence and reducing its ability to control predators. Energy leaks have any number of causes, among them:

▸ A hot wire touching a damp wooden fence post

▸ Vegetation growing against fence wires

▸ Broken tree branches or other debris on wires

▸ Broken, sagging, or poorly connected wires

▸ Too many splices in the line wires

▸ Leaves, dust, bugs, or small creatures lodged in insulators

▸ Cracked or missing insulators

▸ Insufficient grounding

▸ Rusty wire or connectors

▸ Salty sea air

▸ Type of fence wire used

▸ Fence too long for size of energizer

Most of these situations cause arcing. Pulses of current, instead of flowing smoothly from one place to another, jump across a narrow gap by means of a spark. When arcing occurs, you can usually hear the snap of the jumping spark, which can help you find and correct the source of leakage. If you hear snapping but can't find the short, go out at night and you'll actually see the spark.

Predators constantly test your fence, and so should you. Check it, too, after every lightning storm or deep snowfall, and any time you make a change or repair. Note the normal voltage of your fence and check it periodically — at least weekly, preferably daily — for any changes. Checking your fence should be a regular part of your poultry keeping routine.

rain runs off the roof, or where the grass is greenest. During really dry spells you may have to periodically soak the soil surrounding the rods to keep your fence working properly. But don't place rods directly in a running stream or along a stream bank, or you could end up with a dead fence and an electrified stream.

Provide a minimum of 3 feet of rod in the ground per joule of your energizer's output. Place each rod close to a fence post, where it won't be in the way when you mow, and connect all the nearby ground wires to the rod. For safety's sake, keep rods at least 50 feet from the nearest power pole or other grounded utility. Position the first rod as near as possible to the energizer and space additional rods at least 10 feet apart. Rods that are closer together than 10 feet will function as one, wasting your grounding potential.

For a wire-return fence that's more than 1,000 feet long, space rods every 1,000 to 1,500 feet if your fence is short and/or conditions are typically dry. If your fence is long and your soil is reasonably moist, you can get by with rods spaced every 3,000 to 5,000 feet. For a fence that includes steel T-posts, grounding can be enhanced by using steel T-post clips to attach the ground wires to the steel posts.

Electrified Plastic Netting

Electrified plastic net fence, or electroplastic netting, has become popular with urban chicken keepers. It consists of electroplastic twine (also known as polywire) woven into a net mesh and comes completely preassembled with plastic or fiberglass posts and hot and ground connector clips. Support ends and corners are tied with guy lines secured with tent pegs, both of which are included.

Cost varies with brand, height, and mesh size. The tallest, in the 48-inch range, may be used for most types of poultry and to deter foxes, coyotes, dogs, and cougar. The middle height, in the 40-inch range, is used to confine nonflying ducks and geese and to keep out cats and skunks. The shortest, about 30 inches, is low enough for most poultry keepers to step over and may be used to confine weeder geese and turkeys and to control raccoons and mink.

Suppliers of electroplastic net fences usually recommend the type of energizer to use with their brand of fence. You could use a larger energizer without harming the net, but too large an energizer could prove fatal to poultry and other small animals.

This simple system sounds great in principle, but in practice it is not ideal. It must be constantly electrified so poultry, pets, and predators won't get tangled up and strangle in the net. If you live in an area susceptible to power outages, you must use a battery- or solar-operated energizer to make certain the fence is always fully functional. Even when the fence is hot, poultry may get tangled in the net and electrocute themselves, in the process tearing the net and thus reducing its useful life.

Other issues — difficulty keeping the net from sagging; problems getting line posts into rocky soil, drought-ridden clay, or frozen ground; the need to anchor the fence in windy conditions; guy wires at the corners to trip over; having to constantly monitor grass and weed growth; and the probability that young birds and smaller poultry breeds won't respect the fence — can become incidental in comparison to finding dead bullfrogs, turtles, and other wildlife trapped in the electrified net.

The incident that most discouraged me from using this type of fence was finding one of my roosters tangled in the net and getting repeatedly defibrillated. Although I nearly had a heart attack myself, the rooster survived, but only because his rescue was timely. Electroplastic net fence may work okay for small flocks of a sedate breed or species but can be hazardous for such active poultry as guinea fowl that constantly chase one another or geese that jostle aggressively during breeding season. Its best use is as a portable fence that lets you periodically move your flock to fresh forage.

Stock Panels

A relatively fast yet secure way to fence a poultry yard is to use livestock panels, also called feedlot panels, rigid wire panels, stockade panels, or simply stock panels. Any sizable farm store carries them or can order them. Each panel is made of super heavy galvanized wire rods, welded into sturdy 16-foot-long sections.

Heights vary, depending on the type of livestock the panels are designed to contain. None is made specifically for poultry,

but several options are suitable for the purpose.

A hog panel is only 34 inches high and commonly has horizontal rods closer together at the bottom (typically 2 inches) and gradually increasing to 6 inches apart at the top; vertical rods are typically 8 inches apart. A cattle panel is 50 or 52 inches high and usually also has vertical rods 8 inches apart and horizontals that start 4 inches apart at the bottom and increase to 6 inches apart at the top. The cattle panel is a better height for poultry, but the hog panel has more secure bottom spacing. A combination panel solves both problems; it is tightly spaced at the bottom like a hog panel but has the height of a cattle panel.

Where greater security might be needed, use horse panels or utility panels to make a taller fence with tighter mesh. Horse panels are welded with a 2- by 4-inch mesh throughout, and some are as high as 5 feet. Utility panels likewise have the same mesh size throughout, usually 4- by 4-inch, and are available in heights from 5 feet to as much as 8 feet. Utility panels are longer than other stock panels, coming in 20-foot lengths, making them more difficult to maneuver.

Because these sturdy panels are designed for heavy livestock, they won't sag or collapse if pushed into or climbed on by a large predator. They are easier to erect than other types of fence because they don't have to be stretched, and you need only two posts per panel. They may be attached to wooden posts using fence staples, or to T-posts using T-post clips. For a maintenance-free permanent fence, weld the panels to steel posts. A panel fence easily accommodates the addition of scare wires on the outside and on top, which lets you get away with using panels of lesser height.

For maximum security, the panels themselves may be electrified — an option suitable only where visitors, especially children, are not likely to encounter the hot fence. For a fully electrified panel fence, use insulators to attach the panels to fence posts. Or use self-insulating fiberglass posts, available from sources offering a full range of electric fence supplies. Fiberglass posts are easier to install than other types of post because they don't require the muscle needed to use a post-hole digger or a T-post driver but instead may be pounded into the ground with a heavy hammer. The folks at Kencove (*www.kencove.com*) suggest using 1¼-inch FiberRod posts, cut in 6-foot lengths and set 2 feet into the ground, for a fence constructed of 50-inch or 52-inch panels.

Fully electrified panels must be raised slightly off the ground to avoid shorting out the fence but remain low enough to keep small predators from slinking underneath. Pieces of 2×4 lumber make good spacers for getting the panels all the same height off the ground before attaching them to the FiberRods. I like to space the bottom high enough to accommodate the deck of my riding mower, so we lay the narrow side of the 2×4 against the ground. Other people place the wide side against the ground, making the fence lower and therefore more secure against weasels and other small predators, but also

requiring more diligent removal of weeds along the fence's bottom.

As each panel is adjusted for height, use a black marker to mark the fiberglass posts where the third horizontal rod from the top of the panel (12 inches from top) crosses the post. At each mark, drill a ¼-inch hole through the center of the post and, from the enclosure side of the fence, insert a ¼-inch-diameter, 2½-inch-long stainless-steel bolt that is threaded for at least 1½ inches. Slip on a washer and nut, and tighten the nut snug to the post. Slip on a larger, 1½-inch-diameter washer and nut (this time either add a dab of Loctite to the nut or use a lock nut) and screw it onto the bolt until the washer is ⅜ inch from the first nut.

The first washer keeps the nut from digging into the post. The second, larger washer keeps the stock panel from sliding off the bolt. Hang each panel by resting the third horizontal from the top between the first nut and the big washer. With the weight of the panels supported by the bolts, use zip ties to secure the panels to the posts. This fence goes up pretty fast. When energized, it is sturdy enough to protect poultry from the largest predators, including bears.

Scare Wires

Electrified scare wires are used to keep animals from rubbing or leaning against a nonelectric fence, or from digging or burrowing under it. A predator touching the scare wire with its nose or other sensitive part will feel a shock. An animal trying to crawl or dig under the fence will get zapped on its back. An animal trying to climb over or through the fence will feel a jolt to its belly.

Insulators called offset or standoff brackets are used to space the electrified wire far enough from the main fence to prevent arcing or stray voltage. Brackets should be placed close together to keep the scare wire from sagging, especially where polywire is used. A scare wire is typically run 8 to 12 inches above the ground around the outside of the fence. As insurance against climbers, especially if your area experiences deep winter snow, add a second scare wire 8 to 10 inches below the top of the fence.

If the main fence is made of wire and is well grounded, stringing a scare wire right above the top will discourage perching — any raptor or other predator that rests against the hot wire will press it down until it touches the grounded wire.

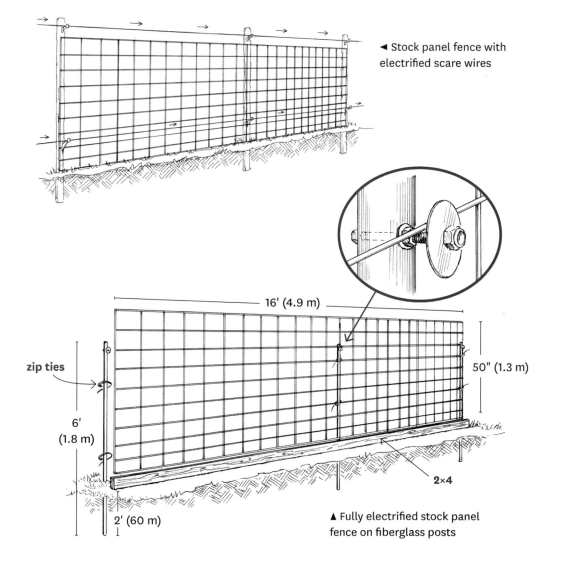

◄ Stock panel fence with electrified scare wires

zip ties

16' (4.9 m)

50" (1.3 m)

6' (1.8 m)

2×4

2' (60 m)

▲ Fully electrified stock panel fence on fiberglass posts

Hot wires along the top may also be used to increase the overall height of a woven wire fence. Run the first strand 2 inches above the woven wire. For more height, make the second strand 6 inches above the first. A third strand could go 8 inches above the second.

Along the driveway to our barn is a short section of rail board fence that ends where the high-tensile electric fence starts. To the top rail we attached a grounded metal strip, and above the strip we ran a strand of electrified polywire. We no longer have to worry about raccoons climbing the fence or raptors resting on top before swooping into the poultry yard. We just have to make sure unsuspecting human visitors don't do the farmer thing of leaning across the top rail while gazing at our chicken flock.

Fence with Electrified Scare Wires

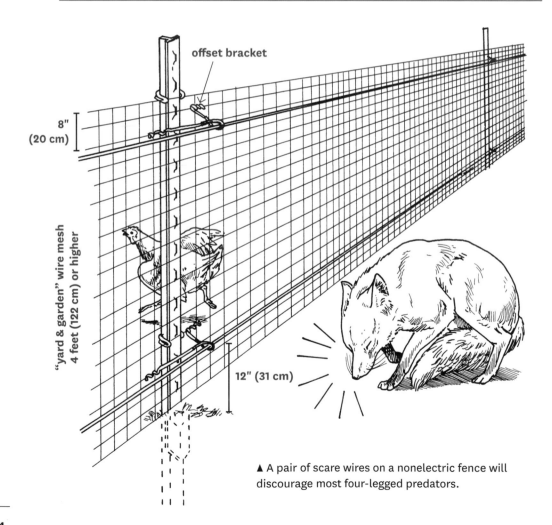

offset bracket

8"
(20 cm)

"yard & garden" wire mesh
4 feet (122 cm) or higher

12" (31 cm)

▲ A pair of scare wires on a nonelectric fence will discourage most four-legged predators.

Good Fences Make Good Neighbors

On our Tennessee farm, we frequently hear coyotes yipping in the dark of night. A wildlife officer once told me that keeping coyotes away from poultry is easy. All you need is a good fence. But should a coyote once find its way through the fence and get a taste of fresh chicken, keeping it from coming back would be extremely difficult, if not impossible.

Accordingly, our main poultry yards are surrounded by multiwire high-tensile electric fences kept hot with high-energy, low-impedance energizers. Patrolling the fence lines and controlling weeds to ensure the fences remain fully functional is a top priority.

One evening while my husband and I were engaged in evening chores, we heard a gang of young coyotes yipping excitedly in the nearby woods, no doubt out celebrating life. The sound got louder as the pups got closer, until suddenly we heard a couple of surprised yelps. And then silence.

We can only guess the pups, intent on their joyful frolic, had failed to notice the fence until their wet little noses engaged with a hot wire. A friend who is knowledgeable about coyotes told me she wouldn't be surprised if Mom and Pop Coyote had taken their pups out for an evening stroll and deliberately let them romp into the hot wire as an object lesson.

Coyotes still regularly patrol our farm. We frequently find their photos on our trail camera and their scat peppering our driveway. But so far (knock on my wooden head!) we haven't seen any sign of coyotes anywhere near the barnyard. The unexpected shock and awe delivered by that powerful electric fence apparently left a strong and lasting impression on our local coyote family.

SECURE GATES

No matter how secure your fence is, it's only as secure as your gates. With both of our professionally installed chain link poultry yards we were left to deal with predator-size gaps at the sides and bottoms of the gates. Even when a gate is initially installed close enough to the ground to exclude predators, traffic from walking, wheelbarrows, mowers, and so forth will eventually wear grooves under the gate. Adding a gate sill will solve the problem.

Any easy method to create a sill is to sink a pressure-treated 4×4 under each walk-through gate, and a 6×6 under a drive-through gate. Or you could pour reinforced concrete sills of similar size. Or you might opt to install Dig Defence gate plates (see the Resources), designed to discourage digging. Whichever option you choose, your investment will prevent soil compression from creating ruts beneath your gates — helping keep your birds from slipping out and predators from slipping in.

Another often-overlooked entry point for predators is a nonelectrified gate in an electrified fence. What's to prevent a predator from climbing or leaping over the gate? The solution we use on our farm is scare wires, one along the lower outside of the gate and another along the top. The gate's scare wires are attached to the fence's hot wires with gate handles made for the purpose. A gate handle has a loop at one end for attaching the gate's scare wire and a hook at the other end to attach to a loop in the fence's hot wire. The handles are insulated so you can unhook them without getting zapped.

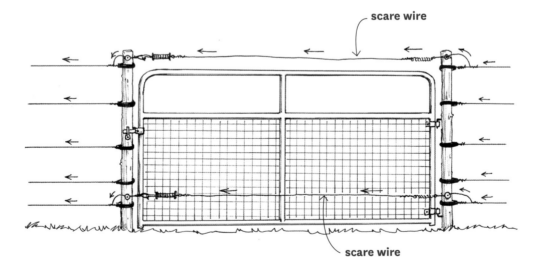

scare wire

scare wire

▲ An often-overlooked point of entry for predators is a nonelectrified gate in an electric fence. Here's how to rectify the situation.

6. THE POULTRY PERSPECTIVE

"You'll get results by always putting yourself in the other fellow's place and thinking about what you would do yourself."

Sherlock Holmes

So far we've examined the methods by which poultry predators operate and how to foil their plans. In this chapter we'll look at things from the poultry perspective, starting with how chickens see in comparison to how predators see. We'll also consider a predator attack from the poultry perspective: How do chickens and other poultry react while being stalked or when caught by a predator?

VISION: PREY VERSUS PREDATOR

Vision is the most important sense for most birds, including poultry. Birds of prey need keen vision for hunting. Birds that are preyed upon need keen vision to avoid predators. Poultry, along with most other prey animals, have eyes on the sides of their heads. By contrast, birds of prey, as well as most other predatory animals including humans, have forward-facing eyes. Predator and prey therefore have different ways of seeing.

▲ Eyes in front, the animal hunts. Eyes on the sides, the animal hides.

How Chickens See

A chicken's eyes, positioned on the sides of its head rather than toward the front, can see two different things at the same time. This monocular vision gives chickens and other backyard poultry a wide panoramic view. The only place where a chicken sees the same thing out of both eyes is directly in front of its beak, a feature that lets it accurately peck a bug or a kernel of grain off the ground.

Because chickens spend most of their time pecking on the ground, the ability to see what's happening along both sides and to the rear without having to turn their heads has the advantage that they can watch out for predators at the same time that they're busy filling their crops. The chief disadvantage is that they have a blind spot straight ahead, which is why a chicken turns its head to see something approach from the front.

Another reason a chicken turns its head for a better view is that, compared to most other animals, birds have large eyes. Relative to the size of its head, a chicken's eyes are about 25 times bigger than a human's eyes. Their eyes don't look so big because they're partly hidden behind feathers. But the eyeballs take up so much space inside the skull that they leave little room for muscles. As a result, a bird's eyes have a limited ability to move within the eye socket. When a chicken wants to focus on something, instead of moving its eyes, it moves its entire head.

Which way the chicken turns its head depends on what it's trying to see. Like other birds, a chicken has a right-eye system and a left-eye system, each with different and complementary capabilities. The left eye transmits information to the right side of the chicken's brain, and the right eye transmits information to the left side of the brain. Unlike human eyes, which move together to both focus on the same thing at the same time, a chicken's eyes move in opposite directions to take in two different scenes at the same time.

The right-eye system is less easily distracted as it focuses on close-up activities, such as seeking food items on the ground. The left-eye system pays more attention to distractions, such as unusual movement that could signal the approach of a predator. By noting the direction and speed of approach, the left-eye system (right side of the brain) decides how the chicken will respond. If the response is fear, the left-eye system seeks an escape route. This dual eye system gives a chicken the ability to sleep with one eye open, resting one side of the brain at a time while the other side maintains vigilance.

At the back of a chicken's eyeball is a focal point, or fovea, that functions as an image enlarger to provide a sharper image when viewing distant objects. Unlike humans, the chicken has a second fovea at the side of the eye to enhance sideways viewing of close-up objects. A chicken trying to get a good look at something will turn and tilt its head for a better look through one or the other foveae, somewhat like a person trying to get used to wearing bifocals for the first time.

Chickens have better color vision than most animals, including humans, because their retinas are organized primarily for seeing during the day, when they spend most of their time looking for things to eat and watching out for predators. Daytime vision relies on light receptors in the eye's retina, called cones, which are most effective in bright light. Cones are essential for detecting color, which helps the bird find tasty food, not to mention helping the bird select healthy mates. The light receptors in the eyes of a chicken, and most other diurnal birds, are about 80 percent cones, compared to a human's meager 5 percent cones.

FOUR FUN FACTS ABOUT BIRD VISION

▶ Birds see colors humans can't see. They see four wavelengths: red, blue, green, and ultraviolet. By comparison, humans see only the first three, so we can only guess at what a bird actually sees.

▶ Birds have color filters in their eyes — tiny droplets of colored oils — that increase the range of colors they see, as well as increasing contrast and brightness.

▶ A bird can see fine detail in objects two to three times farther away than can the average human.

▶ Owls and other nocturnal birds see better in the dark than humans, but we humans see better in the dark than chickens and other diurnal birds.

Chickens also have a double cone receptor, which helps them detect motion better than we humans can. They don't, however, see things well that aren't moving, which is why a wily predator skulks closer only when it believes the chicken is looking elsewhere, and why chickens are so easily startled by any sudden movement.

The trade-off of having lots of cones for superior day vision is that chickens and other backyard poultry don't see well after dark. Night vision relies on receptors called rods, which function better than cones in low light. The chicken has many fewer rods than cones, so it can't see well after dark and therefore becomes more vulnerable to night-hunting predators. It's also why chickens are easier for their human keepers to catch at night than during the day.

How Ducks See

A domestic duck's eyes are positioned higher in the skull than a chicken's eyes, giving the duck a better view of what's overhead. If you look at a duck from the back of its head, you can see its eyes bulging out at the sides. As a result, without

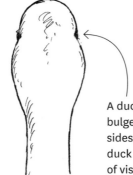

A duck's eyes bulge out at the sides, giving the duck a wide field of vision.

moving its head the duck can see horizontally all the way around, and vertically from above its bill and over the top of its head to more than 180 degrees behind its head.

Like chickens, ducks can sleep with one eye open. In a group of ducks sleeping on the ground, those on the outer edges of the group open the outward-facing eye to watch for predators, while those that are surrounded by other ducks get to sleep with both eyes closed. After a time the outer-sleeping watch ducks turn their bodies in the opposite direction to watch for predators with the other eye while giving the first eye a rest.

How Predators See

Because we predators have eyes toward the front of our heads, we have a greater range of binocular vision than poultry and other prey animals with eyes at the sides of their heads. With binocular vision, both eyes see the same object at the same time, but not precisely from the same angle. To demonstrate this for yourself, hold a finger in front of your face. Now close first one eye and then the other. The finger seems to jump to the right when viewed only with the left eye, and to the left when viewed with the right eye. Your brain combines the two images into one.

Humans and other animals with a wide field of binocularity have really sharp vision in only a narrow field directly ahead. To demonstrate this small area of central vision, make a fist and hold it in front of your face at arm's length. Unless you shift your eyes, everything surrounding your fist appears less sharp than your fist itself,

Comparing Fields of Vision

PREDATORS

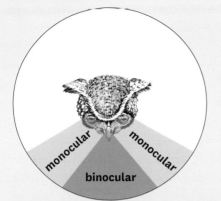

Owl, eyes toward the front of the head:
binocular is 30–60°, total is 120°

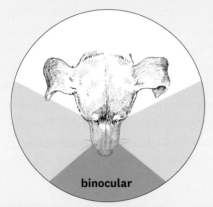

Dog, eyes toward the front of the head:
binocular is 75°, total is 250°

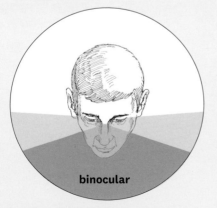

Human, eyes at the front of the face:
binocular is 120°, total is 190°

PREY

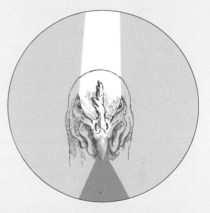

Duck, eyes on the sides of the head:
binocular is 5°, total is 360°

Chicken, eyes on the sides of the head:
binocular is 15–30°, total is 340°

or whatever appears in your central vision straight ahead when you move your fist and see what was behind it. When a cat or a dog sneaks up on something, it does so with its nose pointing directly at its prey. If anything outside the animal's central vision catches its attention, it turns its head to get a better look.

Prey animals generally have a much narrower field of binocular vision, because monocularity is more important for their survival, and they tend to use monocular vision even while hunting. Yet they rely on binocular vision for accuracy in grasping and ripping into their own small prey.

Binocular vision allows depth perception, or the ability to accurately determine how far away things are in relation to where you are, especially when an object is moving toward or away from you. Along with depth perception, binocularity allows spatial localization, or the ability to concentrate on objects at a certain distance while ignoring objects that are closer or farther away. An advantage to spatial localization is that it makes camouflaged creatures visible that would be basically invisible when viewed through one eye.

Camouflage is a survival tactic used by creatures that masterfully blend into their natural surroundings. By concealing a prey animal's location and movement, camouflage allows it to evade predators. And a camouflaged predator can more easily sneak up on prey. The ability of binocular vision to see through camouflage is called "breaking camouflage." It works because the camouflaged creature is closer to the viewer than the creature's background.

To demonstrate breaking camouflage (assuming you see equally well with both eyes), hold up one finger in front of a distant object. First look at the finger, then at the distant object. When you look past the finger into the distance, the finger appears as two fingers, both of which are nearly transparent. In other words, you can see things that are hidden behind your finger — X-ray vision!

Some feather patterns blend into the natural surroundings to provide camouflage.

REACTION TO PREDATORS

Fear of predation can be a significant source of psychological stress for backyard poultry. Flocks get restless at nightfall when they know a predator will be on the prowl. Chickens may pile on top of each other, away from possible entry points of potential predators, smothering those unfortunate enough to be on the bottom. A flock may be unwilling to venture outside to forage during the day, and hens may be reluctant to lay eggs. Chickens and other poultry species have developed several strategies for coping with the possibility of predation.

Eat and Run

Chickens and other backyard poultry forage most actively in the early morning and early evening. At midday, and again at night, they settle somewhere comfortable and safe to digest what they have eaten. Having a crop in which to store large amounts of food lets them eat a lot at once, getting enough to eat while actively foraging, and avoiding being eaten while resting and digesting their recent meal.

Healthy poultry operate at a high body temperature because they have naturally high blood sugar — about twice that of a human — which gives them the energy to keep on the move while foraging. High body heat also increases certain physical functions, such as the transmission of data through the nerve cells, allowing the bird to quickly react to incoming information like the sudden appearance of a predator. Interestingly, the smaller (more vulnerable) breeds have a higher body temperature than the larger breeds, and males (ever vigilant) have a higher body temperature than females.

Circle of Safety

Like all animals, each member of a flock maintains a circle of safety within which it feels protected. The closer a predator comes to the outer edges of the circle of safety, the more threatened the bird feels and the more nervous it gets. The distance a bird keeps between itself and a potential source of danger is variously called critical distance, fright distance, or flight distance and extends vertically (in the case of aerial predators) as well as horizontally (for ground predators). When a predator invades the circle of safety, the bird becomes alarmed and looks for ways to escape.

The breeds with the largest circle of safety — in other words, the high-strung, excitable, so-called flighty breeds — are those best able to evade predators, especially when the flock is free to roam out in the open. Among chickens these breeds include the best layers and the game breeds. Guinea fowl are also good at evading predators, except for hens that are nesting or guineas that persist in roosting overnight outdoors in a tree.

FLIGHTY CHICKEN BREEDS

- ▶ Ancona
- ▶ Andalusian
- ▶ Campine
- ▶ Catalana
- ▶ Chantecler
- ▶ Crevecoeur
- ▶ Empordanesa
- ▶ Fayoumi
- ▶ Hamburg
- ▶ Jungle Fowl
- ▶ Kraienkoppe
- ▶ LaFleche
- ▶ Lakenvelder
- ▶ Leghorn
- ▶ Malay
- ▶ Marans
- ▶ Minorca
- ▶ Norwegian Jaerhon
- ▶ Old English Game
- ▶ Penedesenca
- ▶ Redcap
- ▶ Sicilian Buttercup
- ▶ Spanish
- ▶ Spitzhauben
- ▶ Sumatra

FLIGHTY WATERFOWL

- ▶ African geese
- ▶ Chinese geese
- ▶ Campbell ducks
- ▶ Crested ducks
- ▶ Runner ducks

Dress for Success

In the area where I live, as in many other areas, white breeds of poultry are not popular for backyard flocks because they tend to be the first to get picked off by predators. Much preferred are breeds that are less highly visible. To name but a few: barred Plymouth Rocks, speckled Sussex, black Marans, and Rhode Island Red chickens; pearl, royal purple, bronze, and copper guinea fowl; black, bourbon red, bronze, and Narragansett turkeys; Cayuga, khaki Campbell, mallard, and Rouen ducks; African, brown Chinese, buff, and Toulouse geese.

Similarly, popular strains for free-ranging meat chickens are colored Cornish hybrids, in contrast to the white Cornish broilers favored by industrial poultry producers with their vast indoor production sheds. Most colored strains have red feathers, although some are barred, black, gray, or any color other than white. Colored Cornish hybrids go by a variety of colorful names such as Freedom Ranger, Red Ranger, Black Broiler, and Silver Cross.

The breeds and strains named here are by no means the only options. Any dark or patterned breed is more predator resistant than a white breed. No matter the color, though, no breed is predator *proof* if not housed within predator-proof facilities.

Feather Features

Guinea fowl have a unique way of dealing with being caught, whether by a predator or by a human handler. If a predator grabs a guinea by its feathers, the guinea will release the feathers, leaving the surprised predator with a mouthful of feathers (or the human with a handful of feathers) while the guinea escapes, albeit perhaps somewhat on the naked side. Many's the time I've found piles of guinea feathers in the yard, even when all my guineas were present and accounted for.

A poultry feather feature that works in a predator's favor is an abundance of head feathers that interfere with a bird's ability to see the predator coming. Some of the chicken breeds have extreme topknot feathering that hangs forward into the birds' eyes. Polish, Silkies, and other crested breeds are less suitable than others for free ranging or other situations that potentially expose them to predators. One spring I lost — on two different days, to an unknown predator — a young pair of visually impaired buff laced Polish.

If a topknot chicken is not destined for the show circuit, clipping back the offending feathers could be life-saving. Sometimes you'll see the suggestion to use a pony tail band or hair clip to hold back the front feathers, but an annoyed chicken can scratch them off with a claw and might be tempted to then consume a worm-like elastic band, resulting in choke or impaction.

Flock Together

An interesting thing happens when different poultry breeds are kept together in the same run: those of like color tend to stick together. On our farm, when we raise chicks of two different colors, such as Rhode Island Reds and barred Plymouth Rocks, they tend to forage in color groups. When we had both white and pearl guineas, the white ones all perched together at one end of the roost, while the pearls perched at the other end. Birds of a feather, as they say.

How do you suppose a bird knows what color it is? One distinct possibility is that, as a group, red chickens (for example) might shun or chase away a barred chicken, and conversely the barred group shuns any chicken with red feathers. So it's not that each bird knows what color it is, but rather that it chooses to hang with the gang that makes it feel safest.

From a predatory standpoint, flocking together creates safety in numbers,

SIZE MATTERS

The larger land predators, such as bobcats and coyotes, prefer the larger poultry specimens, but even the largest raptors are less apt to bother big birds that are too heavy to be carried into the air. Turkeys, geese, and the larger chicken breeds, such as Jersey Giants, are therefore less susceptible to predation by raptors than lighter-weight types of poultry, making them more suitable for free ranging in an area that is well fenced against four-legged predators but open to aerial attack.

Birds of a feather really do tend to flock together.

inasmuch as predators have a tendency to select the most conspicuous prey — the one that stands out as being apart from, or different from, others. In other words, the oddball in the crowd. Birds with similarly colored plumage that forage together make it less convenient for a predator to single out one individual from among several that essentially all look alike.

Sound the Alarm

Guinea fowl and geese are notorious for making a loud racket when they sight a predator. Both species have long been used as farmyard guardians. Chickens are not quite as insistently loud, but they seem to have a wider vocabulary when it comes to alarm calls, although it may be just our human impression because chickens have been studied more than any other poultry.

Chickens use different predator alarm calls depending on whether the perceived threat is possible or imminent and whether the predator is approaching by ground or by air. The caution call, consisting of a few quick notes briefly repeated, is made by a chicken that sees, or thinks it sees, a predator in the distance. It is not particularly loud or insistent and doesn't last long unless the predator becomes a threat.

A more insistent alarm cackle announces the approach of an apparent predator at ground level, or perhaps perched on a fence post or a low branch of a nearby tree. Flock mates may join the cackling while craning to get a better look. The cackling increases in intensity the longer the assumed predator is in sight and may continue for some time after the creature leaves.

When a chicken, usually a rooster, spots an approaching raptor, it sounds an air raid call, whereupon its flock mates run for cover. False alarms occasionally occur, but too many false alarms produce the same result as the boy who cried "Wolf!" Chickens therefore learn to differentiate such things as passing crows, buzzards, and light planes from predatory hawks and eagles.

Besides alerting flock mates to danger, an alarm call draws a predator's attention to the caller. But by causing flock mates to flee, the one sounding the alarm also has the opportunity to take cover by joining the confusion of fleeing comrades.

Run and Hide

Poultry types that are capable of flying will attempt to fly away when approached by a predator. This safety feature is one good reason free-range poultry should not be pinioned (have their wing feathers trimmed so they can't fly). Our guineas, for example, prefer to travel around our farm by walking (or, much of the time, running) on the ground but will fly into a tree or onto the barn roof if they sense danger.

Birds that are too heavy to fly, or otherwise not able or inclined to fly, will look for a place to hide. Many a poultry keeper has experienced the sinking feeling of missing several birds after a predator attack, only to find some of them emerging from various hiding places after they feel safe enough to come out.

Once when I went out at dusk to close the door to our Silkie coop, I discovered a 'possum inside the coop and the Silkies nowhere in sight. Until well after dark my husband and I searched the yard, with flashlights in hand, until we found each tiny black Silkie cowering in its own cleverly concealed hiding place.

When a predator approaches chickens pecking in their yard, or free-ranging around a mobile coop, the chickens will run inside, provided they have quick access. Mobile free-range coops are often fitted with wider-than-usual access doors so several birds can get in or out at the same time without causing a traffic jam. In a large yard, or where too many chickens cannot rapidly crowd through a small pophole, establishing cover in the poultry yard can give them places to tuck themselves in until danger passes.

Holler Bloody Murder

Any type of poultry that's nabbed by a predator will squawk loudly, if it is able. Some predators take preemptive action by grabbing a bird by the throat, either choking it or breaking its neck before it can make a sound.

The typical distress squawk consists of a series of loud, long, repeated calls made by a chicken or other backyard bird that's been captured and is being carried away. The intent of the squawking may be to frighten the predator into letting go, but it also serves to warn flock mates of present danger. The typical response of flock mates is to run and hide, although occasionally a courageous bird (usually a male) will try to rescue the one in distress by attacking the predator.

Fight Back

Some types of poultry typically respond to a predator attack by attacking the attacker. A goose or a Muscovy duck will use its powerful wings to bludgeon a predator into letting go. Chickens, and ducks, too, may violently flap their wings, which

An extra-wide chicken door allows multiple chickens to get into the coop in a hurry, but a door that's too high invites hawks, crows, and jays to venture inside as well. An ideal height for most chickens, guineas, ducks, and lighter breeds of turkeys and geese is 15 inches.

can cause a startled predator — especially one that's young and inexperienced — to let go. Geese and turkeys may also attack small predators, such as crows or ground squirrels, that are after eggs or hatchlings.

Among backyard poultry, guineas are unique in their habit of mobbing a predator (see Diurnal Foxes on page 168), a defense that's more common among crows and other wild birds. Cooperative harassment by mobbing is a highly effective way of dealing with a predator. It serves these several purposes:

- It provides a warning to unwary flock mates.

- It blows the predator's cover, destroying any chance of a stealth attack.

- It teaches younger flock members to recognize dangerous species.

- It lures the predator away from nesting sites.

- It has the potential of inflicting injury to the predator.

- It demonstrates the flock's physical fitness and therefore uncatchability.

- It thereby discourages the predator, sending it away hungry.

Play Dead

A chicken or other backyard bird that's grabbed or carried by a predator may go limp, as if dead. The technical term for this defense mechanism is tonic immobility. It's a good defense against dogs and similar predators that rapidly lose interest as soon as their prey stops moving.

It's also a good defense against a predator that fails to maintain focus on its prey, in which case the bird has an opportunity to escape. For example, a fox that nabs a duck incubating a nestful of eggs may set the apparently dead duck aside while it first eats the eggs. The duck, having been let go, takes the opportunity to "come back to life" and flee. Foxes that are experienced hunters learn to kill their prey before turning their attention elsewhere.

Raising a Ruckus

If you are fond of dividing the world into two kinds of people, then you know that some people hate guinea fowl because they are so noisy and some people love guinea fowl because they are so noisy. We keep a flock of guineas, not for their notorious noise making, but for their legendary appetite for insects — especially insects that damage fruit trees or inflict painful bites on humans.

Originally, our sizable flock of guineas lived near the house. But their morning racket served as a too-early alarm clock, so we moved them to a barn distant from the house. Still, they patrol widely across our farm and into the woods, constantly chattering to keep in touch with one another as they spread out to forage.

When guineas encounter anything that makes them nervous, they sound a loud and unmistakable alarm. We frequently hear their alarm whenever our guineas venture into the surrounding forest. This characteristic warning call is what makes guinea fowl so highly valued by many rural poultry keepers, including our neighbor Jim, who produces pastured cattle, hogs, and chickens.

Jim's guineas are the first half of his two-part predator alarm system. Jim tunes out his guineas' daily chatter, but his smart dog doesn't. When the guinea din intensifies, the dog alerts Jim by pricking up his ears and growling, letting Jim know that immediate action is needed. Jim's combination of boisterous guinea fowl and attentive dog ensures the safety of his chickens and other pastured livestock.

PART TWO
THE SUSPECTS

Red-Tailed Hawk

AVERAGE LENGTH: 20 inches (50 cm)
AVERAGE WEIGHT: 2½ pounds (1130 g)
AVERAGE WINGSPAN: 4 feet (120 cm)

With an overall population of some two million, red-tailed hawks (*Buteo jamaicensis*) are the most prevalent daytime predators of poultry throughout North America. They prefer a diet of rodents, rabbits, and other small mammals, although about 10 percent of their diet consists of birds — including crows, pigeons, quail, and songbirds. A hungry red-tail won't hesitate to attack a barnyard bird up to about the size of a stout rooster and can carry off a smaller bird weighing up to about a pound.

A pair of red-tails mates monogamously, often using the same nest for several years. They aggressively defend their territory during breeding season, and sometimes year-round in places where the local hawk population swells during migration. They defend their territory not just against other hawks, but also against eagles and great horned owls, and in doing so protect any nearby poultry yards from predation by these bigger raptors.

During breeding season, while the female guards the nestlings, the male hunts to provide for his family. When they are not cooperating to raise their young, they often hunt in pairs. Their dietary preference for rodents makes them additionally beneficial to the poultry keeper.

A red-tail hunts throughout daylight hours, varying its strategy depending on the available prey. It may perch on a utility pole, atop a fencepost, or in a tree and employ the sit-and-wait technique — appearing disinterested until the target becomes distracted, then swiftly swooping in.

While hunting or guarding against intruders into their territory, red-tails may also circle over an open field, using little energy by catching currents of warm rising air. On a windy day, a red-tail may appear to hover in midair without flapping its wings. When it does flap, the wing beats are slow and deep, with the wings fully extended on the downstroke and the flight feathers widely spread like fingers.

While in the air, a red-tail periodically emits an extended 2- to 3-second scream familiar to anyone who watches movies or television. Filmmakers are fond of applying the red-tail's unique cry to any species of hawk or eagle.

TYPICAL SIGNS. A red-tailed hawk will carry a small or young barnyard bird to a feeding perch, where it will behead and pluck

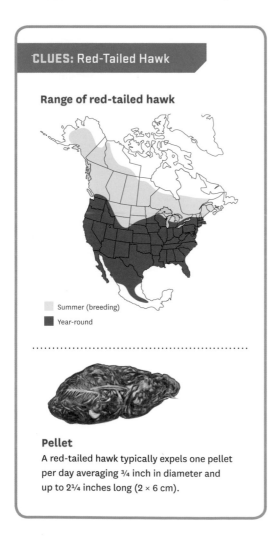

CLUES: Red-Tailed Hawk

Range of red-tailed hawk

Summer (breeding)
Year-round

Pellet
A red-tailed hawk typically expels one pellet per day averaging ¾ inch in diameter and up to 2¼ inches long (2 × 6 cm).

the bird before eating it. The skull and feathers may be found scattered at the base of a tree or utility pole. Unless you locate the feeding perch, the only sign will be a small bird missing during the daytime.

A slightly larger barnyard bird that's too heavy to carry will be plucked and the meat stripped off the bones, usually starting at the neck and working down the breast. The lightened carcass may then be carried to a feeding perch for further snacking. In this case all you'll find is a circle of feathers on the ground. If you look closely at the plucked feathers, you might detect beak marks on the shafts.

A full-size chicken or duck, too heavy to carry even after it's been partially eaten, may be left within the circle of feathers. The hawk may return repeatedly to finish off the carcass or, if the dead bird has been removed, may come back to kill another.

Rarely will a red-tail kill more than one bird at a time. Once, when I saw a dead chicken in my yard and heard a ruckus inside the coop, I expected to find a four-legged predator terrorizing my flock. Instead I saw a hawk attacking a second chicken. To my dismay, a third chicken inside the coop was mortally wounded. This hawk was undoubtedly young and inexperienced, having not yet read the red-tail operating procedures manual.

SEASONALITY. Throughout most of North America, the red-tail female lays her first egg in mid- to late March. She'll lay an egg every other day until the nest accumulates up to about five eggs. She then incubates the eggs for 28 days, and they typically hatch around the end of April. To feed the young and growing brood, the male red-tail must hunt tirelessly.

About 45 days after they hatch, the young birds develop enough feathers to fly. Their parents will continue to feed them for another three weeks. Most juveniles are able to catch their own food within about six weeks of leaving the nest, although their parents may help support them for an additional couple of weeks.

The juveniles start out hunting easy prey — insects and spiders — and move up the food chain as they gain skill. First-year juveniles that lack the experience to catch free-flying wild birds may target free-range poultry, earning the red-tailed hawk the nickname of chicken hawk, a distinction it shares with two other hawk species, the Cooper's hawk and the sharp-shinned hawk.

APPEARANCE. The average weight of a red-tail is about 2½ pounds, with females typically weighing some 25 percent more than males. This difference in size expands a pair's range of prey. Both genders have a short, widely fanned tail and long, wide, rounded wings. The trailing edges of the wings bulge outward at the secondary feathers, and the ends of the wing feathers separate to create the appearance of fingers.

More than a dozen subspecies have regionally variable plumage color, with some forms being considerably darker and some considerably lighter. Commonly the color is brown, darker on top and paler underneath. Dark feather tips on the undersides of the wings form an outline along each wing's outer edge. Most red-tails have a dark band across the breast and red feathers on top of the tail, with lighter tan or pinkish feathers underneath. Although the red-tailed hawk gets its name from its distinctly brick- or cinnamon-red tail, juveniles — which are much slenderer than their chunky parents — don't acquire the characteristic red tail until 2 years of age.

HABITAT AND RANGE. Red-tailed hawks are highly adaptable, even in rapidly changing environments. They are more likely than other hawk species to nest in wooded residential areas. Each breeding pair defends a territory of about a square mile.

Red-tails range from central Alaska south through the United States and into Mexico, settling wherever habitat offers open fields and scattered trees or other perching sites for hunting and tall trees or other structures for nesting. They generally shun areas with large expanses of dense forest or, conversely, treeless terrain.

Species living in the southern United States remain residents year-round. Species that breed in Alaska, Canada, or the northern Great Plains generally migrate south for the winter. When their somewhat larger northern cousins muscle in on southern red-tails, the increased population puts additional pressure on the local rodent population, which in turn increases pressure on poultry yards.

BEST DETERRENTS. Red-tails hunt mostly from an elevated perch. Temptation may be reduced by eliminating potential perching sites or making them unattractive to land on. Because a hawk hunts by sight, it is attracted to clearly visible poultry, especially free-ranging birds. Providing ample places — such as trees, shrubs, or range shelters — for poultry to take cover helps make them less visible. A chicken taking a dust bath is particularly tempting because it appears to be injured and therefore easy prey.

The best way to minimize red-tailed hawk predation is to keep poultry inside a covered run. Where the yard is too large to cover entirely, crisscross the top with rope, wire, or fish line and hang shiny reflective objects such as CDs or disposable aluminum pie tins. A flock of guinea fowl, a mature rooster, or a livestock guardian dog can help alert a free-ranging flock of a hawk's approach.

Ranging & Raptors: Rewards vs. Risks

Kathy Shea Mormino, Connecticut, author of *The Chicken Chick's Guide to Backyard Chickens*

Early in my chicken-keeping adventures, I learned the hard way about the difference between chicken wire and hardware cloth. I was less than two months into chicken keeping when a red-tailed hawk reached his razor-sharp talons through the chicken wire surrounding the run, killing one of my Silkie teenagers in broad daylight.

However, I prefer that my chickens live their lives unconfined during the day, so each morning I open the door to the run and let my flock free-range on our half-acre backyard. Unconstrained ranging is a more natural experience for chickens, and they are healthier when dining alfresco while simultaneously getting plenty of valuable exercise. I have instituted numerous layers of protection for my flock with a variety of predators in mind, including fencing the perimeter of the property with hot wire, but none is foolproof. Occasionally one of my flock members is claimed back into the food chain, and I grudgingly accept that fate.

And so it happened that one afternoon through our living room window, I saw a red-tailed hawk standing on top of my frizzled bantam Cochin hen, Phoebe. I rushed outside, arms flailing, shouting like a lunatic, which frightened the hawk away with empty talons. My tiny Phoebe lay motionless on the ground beneath a tree limb upon which sat a plastic decoy owl intended to intimidate hawks. Having suffered several puncture wounds to her sides, Phoebe was in shock, but alive. She went on to live many more years, but other hens have not been as fortunate.

Despite the heartbreak of loss to a predator, I wholeheartedly believe the highest quality of life I can offer my chickens includes free-ranging. I am always mindful that we encroached on the home of wildlife when we built our house in their backyard, not the other way around. There is no correct answer to the question, 'What is the best quality of life I can afford my chickens given my risk tolerance for predators?' Anyone who keeps poultry must weigh the benefits and risks of free-ranging for themselves.

Comparing Buteos and Eagles in Flight

Buteos are hawks with broad, rounded wings, relatively short tails, and soaring flight. Eagles have similar silhouettes, but the birds are much larger and heavier than buteos.

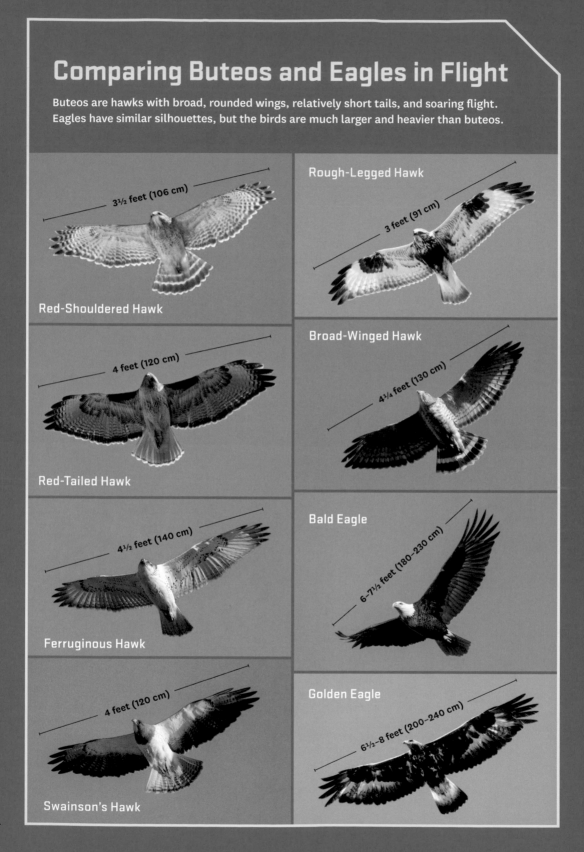

3½ feet (106 cm)

Red-Shouldered Hawk

Rough-Legged Hawk

3 feet (91 cm)

4 feet (120 cm)

Red-Tailed Hawk

Broad-Winged Hawk

4¼ feet (130 cm)

4½ feet (140 cm)

Ferruginous Hawk

Bald Eagle

6–7½ feet (180–230 cm)

4 feet (120 cm)

Swainson's Hawk

Golden Eagle

6½–8 feet (200–240 cm)

Other Buteos

Buteos are the type of hawk described in the song "Oklahoma" — "Ev'ry night my honey lamb and I sit alone and talk and watch a hawk makin' lazy circles in the sky." Although the bird that inspired this image was most likely a red-tail, it might well have been one of the other buteos.

The word *buteo* is Latin for buzzard. In Europe buteos are called buzzards, which can be confusing if you're reading up on buzzards and can't determine if the writer is referring to European hawks or North American vultures.

Because buteos are the most numerous raptors, and because they tend to circle above poultry yards searching for rodents, they are often falsely accused of targeting poultry. That doesn't mean, however, that they aren't occasionally tempted.

Six buteo species seasonally migrate northward into limited areas of states along the Mexican border. Six others, including red-tails, pose a potential threat to poultry throughout parts of the United States and Canada. Three of these species are seasonal — two appearing in the continental United States only in summer, and one only in winter — while the other three remain on the continent year-round.

The five species aside from red-tails are described here in order from the largest to the smallest — size being a direct reflection of the size of backyard poultry that might be of interest. Signs and deterrents are the same as for red-tailed hawks.

▲ The pellet of a buteo, in this case a red-shouldered hawk, is more likely to contain rodent bones than chicken bones.

Ferruginous Hawk

AVERAGE LENGTH: 24 inches (63 cm)
AVERAGE WEIGHT: 3½ pounds (1.5 kg)
AVERAGE WINGSPAN: 4½ feet (140 cm)

Although ferruginous hawks (*Buteo regalis*) are the largest buteos, they rarely bother poultry. They are much more interested in rabbits and rodents, especially prairie dogs and ground squirrels. Because of their penchant for the latter, they have long been considered the poultry keeper's friend, although with changing habitat they have been known to adapt to lizards and birds, including chickens. They hunt in pairs or in cooperative groups of half a dozen or more.

The word *ferruginous* comes from their rust-colored plumage (from the Latin word *ferrum,* meaning iron, and in this case referring to rusty iron), which covers the shoulders and legs. This buteo is one of only two North American hawks to have feathered legs (the other being the rough-legged hawk). The head is gray and the underside is white.

Range of ferruginous hawk

- Summer (breeding)
- Winter (nonbreeding)
- Year-round

Rough-Legged Hawk

AVERAGE LENGTH: 20 inches (50 cm)

AVERAGE WEIGHT: 2¼ pounds (1 kg)

AVERAGE WINGSPAN: 4¼ feet (130 cm)

Arctic dwellers for most of the year, rough-legged hawks (*Buteo lagopus*) move in late fall to southern Canada and the northern United States and remain there until early spring. Their name comes from their leg feathering, a feature that helps them keep warm during arctic winters.

Rough-legs seek winter habitat that is similar to their breeding grounds — open grasslands, fields, prairies, wet meadows, marshes, deserts, and even airports — where rodents are abundant. Although they eat mostly rodents (lemmings are their delicacy of choice in the Arctic), they also eat ground squirrels, gophers, and birds. They typically search for prey by scanning the ground while facing into the wind and hovering in the air — a technique known as kiting. At one time they were misunderstood as a serious threat to poultry, but they much prefer to rid the poultry area of rodents.

Swainson's Hawk

AVERAGE LENGTH:
20 inches (50 cm)

AVERAGE WEIGHT:
2 pounds (1 kg)

AVERAGE WINGSPAN:
4 feet (120 cm)

Although similar in size to a red-tail, a Swainson's hawk (*Buteo swainsoni*) has longer wings; in fact, when it perches, its wings are so long that they extend beyond the bird's tail. The wings are narrower than those of other North American buteos, and the body slimmer. Plumage color is regionally variable, tending to be dark brown above with a lighter brown chest and a white belly, sometimes with a little rust thrown in.

These hawks inhabit the Great Plains of the United States and Canada, favoring open prairie with scattered trees for nesting. In autumn the entire Swainson's population migrates to South America. Expect to see these hawks disappear from September to November and begin to reappear from February to May.

During the summer they eat ground squirrels and other rodents, rabbits, reptiles, insects, and small birds including baby poultry. During the winter, however, they eat mostly insects — particularly favoring grasshoppers, which gives this buteo its nickname of grasshopper hawk.

Range of rough-legged hawk

■ Migration
■ Summer (breeding)
■ Winter (nonbreeding)

Range of Swainson's hawk

■ Migration
■ Summer (breeding)

Red-Shouldered Hawk

AVERAGE LENGTH: 20 inches (50 cm)
AVERAGE WEIGHT: 1½ pounds (6.8 kg)
AVERAGE WINGSPAN: 3½ feet (106 cm)

Favoring woodlands with tall trees near a river or swamp, red-shouldered hawks (*Buteo lineatus*) inhabit the eastern half of the United States and Canada, and California west of the Sierras. They are not shy about nesting in wooded residential areas. They will not tolerate great horned owls within their territory, making them the poultry keeper's ally.

The red-shoulder is a sit-and-wait hunter, typically perching high in the forest canopy and then swiftly swooping down to nab its prey. Its diet consists of earthworms and other invertebrates, amphibians and reptiles, rodents and other small mammals, and birds up to about the size of a dove. Given the opportunity, they may prey on small poultry while searching for rodents and therefore are sometimes referred to as hen hawks.

Broad-Winged Hawk

AVERAGE LENGTH:
15 inches (38 cm)

AVERAGE WEIGHT:
1 pound (4.5 kg)

AVERAGE WINGSPAN:
3 feet (91 cm)

The smallest North American buteo, the broad-winged hawk (*Buteo platypterus*) is, along with the Swainson's hawk, the most migratory. These hawks breed in the eastern half of the United States and southern Canada from about April through September and then head to South America in a mass migration.

This compact, crow-size buteo has a rather large head, stout chest, and short tail. It is easily distinguished from other buteos by its especially broad and pointed wings. It soars in tighter circles than other raptors, and its wing beats look stiff and choppy. This buteo's call is unlike that of other hawks, consisting of a high-pitched two-note whistle.

While hunting, the broadwing perches in a tree or on a utility pole until it spots prey and then swoops down and carries off its prize. It prefers amphibians and reptiles but will eat mice and other small mammals, large insects, and small birds, including juvenile poultry that might be scratching around under a forest canopy or near the forest's edge during spring and summer.

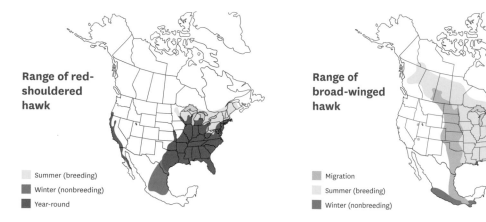

Range of red-shouldered hawk

Summer (breeding)
Winter (nonbreeding)
Year-round

Range of broad-winged hawk

Migration
Summer (breeding)
Winter (nonbreeding)

Common black hawk

Zone-tailed hawk

Buteos in Parts of the Southern U.S.

NAME	LOCATIONS	TRAITS
Common black hawk (*Buteogallus anthracinus*)	AZ, NM, UT, TX	Large, black; favors remote wooded streams; prefers fish, frogs, lizards, snakes
Harris's hawk (*Parabuteo unicinctus*)	AZ, NM, TX	Medium-large, brown and rust; favors desert and urban areas; hunts in teams; eats lizards, rabbits, ground squirrels, medium-size birds
Gray hawk (*Buteo plagiatus*)	southern AZ, TX	Medium size, pale gray, accipiter-like; favors mesquite, cottonwoods, and willows along streams; prefers lizards, but eats small birds up to dove size
Short-tailed hawk (*Buteo brachyurus*)	southern tip of FL	Small, dark brown; flies like a vulture; favors swamp forests, marshes, and suburban woodlands; eats small to medium-size birds
White-tailed hawk (*Geranoaetus albicaudatus*)	southern tip of TX	Medium size, gray with white underside and black banded tail; favors arid grassland; eats rodents, rabbits, snakes, birds up to mallard size
Zone-tailed hawk (*Buteo albonotatus*)	AZ, NM, TX	Medium size, black; highly mimics turkey vultures; favors arid, semi-open woodland; eats ground squirrels, reptiles, birds including young poultry

Accipiters

Accipiters are a genus of hawk distinguished by short, broad wings and long rudder-like tails — features that allow rapid maneuvering through the deep woods. They appear to be in constant motion, with a typical flight pattern consisting of three to six choppy wing beats alternating with a brief glide. These woodland hawks are much more secretive than buteos, and therefore less often seen. *Accipiter*, in Latin, is the generic word for raptor.

Much less numerous than buteos, accipiters have only three species in North America, all of which enjoy a diet of birds. As a group, they are sometimes called bird hawks. Although the largest is the most problematic for poultry, the smaller two share the red-tail's nickname, chicken hawk. Mature accipiters are described here from largest to smallest. Signs and deterrents are the same as for red-tailed hawks.

Accipter   **Buteo**

Northern Goshawk

AVERAGE LENGTH: 23 inches (58 cm)
AVERAGE WEIGHT: 2¼ pounds (1 kg)
AVERAGE WINGSPAN: 3½ feet (107 cm)

The biggest and fiercest accipiters, northern goshawks (*Accipiter gentilis*) are also the least common. They nest in large tracts of mature forest in Alaska, Canada, the northern states, and mountainous regions of the West. Most goshawks are year-round residents, although northernmost pairs may winter in the Great Plains. About every 10 years, when abundant

prey in their home range declines, large numbers of goshawks invade the Southwest.

Goshawks are brown-gray with a striking streaked, barred, or herringbone pattern. They are not quite as big as red-tailed hawks but are just as likely to be attracted to poultry, especially those free-ranging in an open area along the edge of a woodland. They are stop-and-go hunters, perching up high for a short time to survey the area, then gliding among the trees before perching again. When they spot prey, they hurtle down to pounce, talons first. They will kill owls and other hawks infringing on their nesting site and even attack people perceived as a threat.

The word goshawk (pronounced goss-hawk) derives from the Old English words *gos*, meaning goose, and *hafoc*, meaning hawk. Although goshawks also relish rodents and rabbits, they mostly enjoy a wide range of birds including grouse, crows, and poultry up to the size of a duck or, yes, a small goose. They are aggressive hunters with feet and beaks strong enough to take rather large prey.

Range of northern goshawk

■ Year-round
■ Winter (nonbreeding)
■ Winter (scarce)

Cooper's Hawk

AVERAGE LENGTH: 16 inches (40 cm)
AVERAGE WEIGHT: 14 ounces (400 g)
AVERAGE WINGSPAN: 2½ feet (80 cm)

Thriving in forests and woodlots throughout the continental United States into southern Canada, Cooper's hawks (*Accipiter cooperii*) are the most widespread accipiter. They are primarily year-round residents, except that the northernmost among them move to warmer climes during winter. Cooper's hawks are about the size of a crow, with blue-gray plumage on top, red-and-white bars underneath, and wide dark bands across the tail.

These accipiters will eat small rodents, including ground squirrels, and sometimes reptiles and insects but have an enormous appetite for birds. They often take advantage of backyard bird feeders, where feathered snacks are abundant, and are not shy about moving into urban and suburban areas where they find no shortage of pigeons and doves.

Although, compared to goshawks, Cooper's hawks are less troublesome for poultry, they are nonetheless often called chicken hawks — they will take a chicken or any young backyard bird that's easy to nab. Typically they eat the head first, and then the viscera, before starting on the muscles. Like the goshawk, they are stop-and-go hunters. They also have a technique of hiding behind a natural feature or man-made structure while flying close to the ground and then suddenly swooping over the barrier to snatch an unsuspecting bird on the other side.

Sharp-Shinned Hawk

AVERAGE LENGTH: 12 inches (30 cm)
AVERAGE WEIGHT: 5¼ ounces (150 g)
AVERAGE WINGSPAN: 2 feet (60 cm)

Although they are the smallest North American hawks, sharp-shinned hawks (*Accipiter striatus*) make up for their small size with feistiness and stealth. They use the same sneak-attack technique as Cooper's hawks. Unlike other raptors that swoop down on prey, sharpies often fly low along ground contours, where they are less easily detected by an unsuspecting songbird. They are sometimes called chicken hawks, even though they are too small to take mature poultry with any regularity. But they do relish chicks.

Range of Cooper's hawk

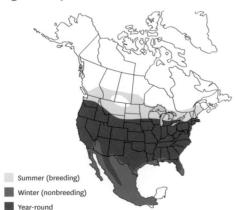

- Summer (breeding)
- Winter (nonbreeding)
- Year-round

Range of sharp-shinned hawk

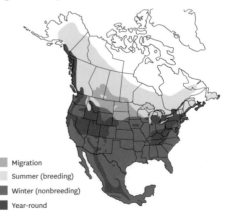

- Migration
- Summer (breeding)
- Winter (nonbreeding)
- Year-round

Cooper's hawk

Sharp-shinned hawk

Cooper's Hawk and Sharp-Shinned Hawk: A Closer Look

FEATURE	COOPER'S HAWK	SHARP-SHINNED HAWK
Size	crow	jay
Shape	evenly tapered	more V-shaped
Head	larger, blocky	smaller, rounded
Head plumage	dark cap on top	hooded appearance
Head in flight	easily visible	barely visible
Tail	longer	shorter
Tip of tail	rounded	squared off
Wing beats	less frequent, less rapid	more frequent, more rapid

HAWKS AND EAGLES

143

Eagles

Although eagles are members of the same family (Accipitridae) as hawks, the two eagle species resident in continental North America are not closely related to each other. The bald eagle is classified as a sea eagle and is of the genus *Haliaeetus*. The golden eagle is classified as a land eagle and is of the genus *Aquila*.

Like hawks, eagles have long, broad wings and a wide wingspan, and they soar high in the sky. An eagle looks so much like an oversize buteo that the two are often confused. Eagles and hawks in North America may be distinguished in the following ways:

- The weight of an average eagle is about three times that of a large hawk.

- The eagle's wingspan is almost double that of the largest hawk.

- An eagle's head and beak are heavier than a hawk's.

- Eagles have stouter feet than hawks.

- Eagles are more likely than hawks to eat carrion.

- Eagles are more powerful than hawks and can take down larger prey.

- An eagle can carry away a chicken or duck; a hawk is likely to eat mature poultry on the spot.

Deterrents are the same for eagles as for red-tailed hawks. Additionally, eagles are more easily discouraged from hanging around by activity and noise created by humans.

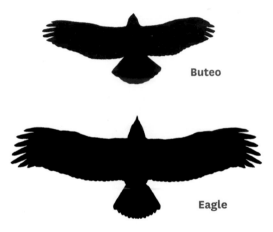

Buteo

Eagle

STATE ANIMALS

The peregrine falcon *(Falco peregrinus)* is the official state raptor of Idaho, the only state to adopt an official raptor.

The bald eagle (*Haliaeetus leucocephalus*), sometimes called the American eagle, became the national emblem of the United States in 1782, against the advice of Benjamin Franklin, who called it "a Bird of bad moral Character" because of its habit of stealing food from other predators. The bald eagle represents great strength, long life, and freedom.

Peregrine falcon

An eagle once polished off one of our guinea fowl in the short time between dawn, when the coop door opens, and when we got there for morning chores, leaving nothing but feathers and bare bones. But poultry predation by bald eagles is rare. In areas where both bald eagles and golden eagles occur, bald eagles are often blamed for the doings of golden eagles.

Bald Eagle

AVERAGE LENGTH: 2¾ feet (84 cm)
AVERAGE WEIGHT: 10½ pounds (4.8 kg)
AVERAGE WINGSPAN: 6½ feet (2 m)

Bald eagles (*Haliaeetus leucocephalus*) are the only eagle species native to North America and nowhere else.

Most bald eagles nest in secluded, timbered areas of Alaska and Canada during summer and spend winters scattered throughout the United States, away from heavily developed areas. Some populations are year-round residents, mostly along coastlines where open water doesn't freeze in winter. During the breeding season, bald eagles are solitary birds, but during winter they may congregate near a major river or lake, where they roost and feed in groups of 100 or more. While migrating, an eagle will eat whatever is easiest to find, which may well be a chicken or a duck.

Since an eagle is big enough to carry off a chicken or a duck, you may find no evidence. On the other hand, you just might find a denuded skeleton in or near the poultry yard.

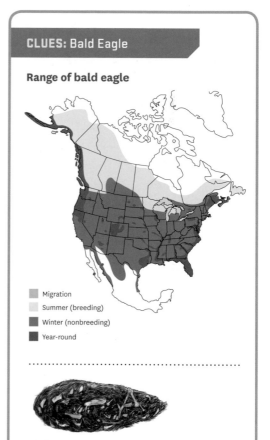

CLUES: Bald Eagle

Range of bald eagle

- Migration
- Summer (breeding)
- Winter (nonbreeding)
- Year-round

Pellet
A bald eagle tends to cast pellets while roosting, sometimes casting multiple pellets after a huge meal that takes several days to digest.

They prefer open terrain — such as desert, plains, rangeland, or tundra — and avoid both highly developed areas and continuous tracts of forest. Although the majority of the golden eagles' diet consists of mammals, they also like large birds, including pheasant-size poultry as well as turkeys and geese.

Golden Eagle

AVERAGE LENGTH: 2½ feet (77 cm)
AVERAGE WEIGHT: 10 pounds (4.5 kg)
AVERAGE WINGSPAN: 6½ feet (2 m)

Although golden eagles (*Aquila chrysaetos*) are nearly the same average size as bald eagles, they differ in several ways. They inhabit open areas and avoid large stretches of forest. They are hunters more than scavengers. They have feathered legs and longer tails. Their color when mature is mostly dark brown with a golden hood (crown and nape feathers). Their genus name, *Aquila*, is the Latin word for eagle. The species name, *chrysaetos*, derives from the Greek words *chrysos*, meaning golden, and *aetos*, meaning eagle.

Golden eagles are much more widespread than bald eagles, being native not only to North America but also to Europe, North Africa, and Asia. However, they are rarely seen in the eastern half of the United States. In the West, southernmost pairs are year-round residents. Northern pairs migrate southward for the winter but, being much more solitary, do not congregate like bald eagles.

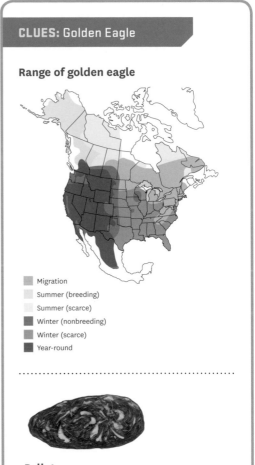

CLUES: Golden Eagle

Range of golden eagle

- Migration
- Summer (breeding)
- Summer (scarce)
- Winter (nonbreeding)
- Winter (scarce)
- Year-round

Pellet
A golden eagle typically casts one daily pellet of similar size to a bald eagle's — as large as 1½ inches in diameter and 3¾ inches long (9.5 × 3.8 cm).

EAGLE OR VULTURE?

Distinguishing a soaring eagle from a soaring vulture can be difficult. Turkey vultures, immature bald eagles, and golden eagles are all big, dark birds that soar and glide. From a distance they look pretty much alike, the more so when eagles join up with a group of vultures. Here are the differences to watch for:

▶ The vulture's wingspan is about a foot less than that of an eagle.

▶ Vultures rock from side to side as they soar; eagles do not.

▶ Vultures soar with their wings raised in a V shape, and golden eagles less so; bald eagles soar with their wings held out flat.

▶ The vulture's wings are narrower and the sides more parallel than the eagle's.

▶ The turkey vulture has a featherless head; a mature bald eagle has a white head; a mature golden eagle has a brown head with a gold hood.

▶ Vultures are more social; eagles tend to be more solitary.

▶ Turkey vultures make little sound; golden eagles are basically quiet but sometimes bark like a small dog; bald eagles are more vocal, making a series of high-pitched piping sounds.

▶ Vultures eat only carrion; bald eagles eat mostly fish; golden eagles eat mostly live prey.

turkey vulture · black vulture · golden eagle · bald eagle

Vultures and eagles all fly with their wingtip feathers spread like fingers. The turkey vulture flies with wings raised in a V, using infrequent, lazy wing beats. Both the black vulture and the golden eagle soar with wings slightly raised, the black vulture flying with snappy, frequent wing beats and the golden eagle with shallow, steady wing beats. The bald eagle flies with wings held straight out to the side and deep, steady wing beats.

turkey vulture

black vulture

golden eagle

bald eagle

Falcons

Small to medium-size raptors, falcons have long tails and long, narrow, pointed wings that make them fast and acrobatic fliers. The word *falcon* derives from the Latin *falx*, meaning curved blade or sickle, which may refer to the shape of the wings in flight, to the talons, or to the beak.

Falcons differ from hawks in several ways. Among their differences:

- Falcons are generally smaller than hawks but have proportionally larger heads.

- Male and female falcons are nearly the same size; female hawks are larger than male hawks.

- Falcons have longer, narrower wings than hawks.

- Falcons fly at higher speeds, while hawks tend to soar or glide.

- Falcons have toothed beaks; the hawk's beak is slightly curved and smooth.

- Falcons have thinner toes and less substantial talons than hawks.

- Falcons kill by crushing with their beaks; hawks kill with their talons.

- Falcons generally attack by steeply diving from the air; hawks pounce from a perch.

- Falcons commonly snatch prey in flight; hawks look for prey on the ground.

Five falcon species inhabit North America, all of which eat birds. They are described here from largest to smallest.

Gyrfalcon

AVERAGE LENGTH: 22 inches (56 cm)
AVERAGE WEIGHT: 3½ pounds (1.6 kg)
AVERAGE WINGSPAN: 4 feet (122 cm)

Gyrfalcons live in remote regions of arctic tundra and are primarily permanent residents, although during the nonbreeding winter months some populations disperse into open fields throughout Canada, occasionally reaching the northernmost United States. They prefer a diet of ptarmigan (circumpolar gamebirds related to grouse) but on rare occasions will take domestic poultry including chickens, turkeys, ducks, and geese.

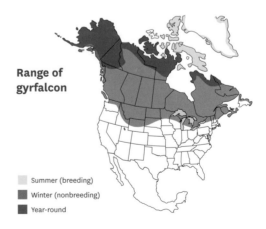

Range of gyrfalcon

- Summer (breeding)
- Winter (nonbreeding)
- Year-round

Peregrine Falcon

AVERAGE LENGTH: 18 inches (46 cm)
AVERAGE WEIGHT: 2½ pounds (1.1 kg)
AVERAGE WINGSPAN: 3½ feet (105 cm)

One of the most widely distributed wild bird species, peregrine falcons (*Falco peregrinus*) are found on all continents except Antarctica. Their plumage is blue-gray on top with pale barring underneath. The largest of the falcons commonly found throughout North America, peregrines are about 75 percent the size of gyrs.

They live in open habitat that provides good hunting, mainly along coastal areas, and are often seen in cities feasting on pigeons and doves. They prefer to catch prey in midair and during an aerial attack can dive at speeds faster than 200 miles an hour. Although sometimes called duck hawks, and capable of taking birds as large as a small goose, peregrines have little interest in domestic poultry that don't typically fly.

Prairie Falcon

AVERAGE LENGTH: 16 inches (42 cm)
AVERAGE WEIGHT: 1½ pounds (686 g)
AVERAGE WINGSPAN: 3¼ feet (101 cm)

Prairie falcons (*Falco mexicanus*) are similar in size and shape to peregrines but reside only in open, dry areas of the western United States, where they are basically permanent residents. Unlike other falcons, they are not primarily bird hunters but prefer ground squirrels, which they pursue by contour hugging. They also fiercely defend their territory against great horned owls, eagles, and hawks. Where prairie falcons take up residence, nearby poultry keepers are less likely to be bothered by other raptors and by ground squirrels.

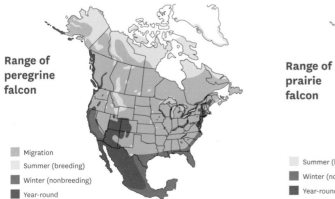

Range of peregrine falcon

- Migration
- Summer (breeding)
- Winter (nonbreeding)
- Year-round

Range of prairie falcon

- Summer (breeding)
- Winter (nonbreeding)
- Year-round

from above, they are more likely to fly along the terrain's contour, frightening birds into taking flight.

They tend to specialize in whatever small birds are most abundant. Even though they may nab an occasional backyard hatchling, their preference for house sparrows — invasive pests that spread parasites and diseases to poultry — makes merlins more ally than enemy.

Merlin

AVERAGE LENGTH: 10 inches (25 cm)
AVERAGE WEIGHT: 7 ounces (200 g)
AVERAGE WINGSPAN: 2 feet (60 cm)

These falcons breed in the northernmost parts of North America. During winter months some populations spread out across the rest of the continent, where they favor environments with open habitat for hunting. In urban areas of the northern prairies, merlins remain year-round residents.

They are considerably smaller than peregrines and have stockier bodies and shorter wings. Like peregrines, merlins prefer to catch their prey on the wing, but instead of diving

American Kestrel

AVERAGE LENGTH: 10 inches (25 cm)
AVERAGE WEIGHT: 4¼ ounces (120 g)
AVERAGE WINGSPAN: 22 inches (56 cm)

A permanent resident of 35 states, the American kestrel (*Falco sparverius*) is the smallest North American falcon and the most widespread throughout the contiguous states. Kestrels are similar in size and shape to mourning doves, and where the two share habitat, a kestrel may be mistaken for a dove. Kestrels are also similar enough to merlins

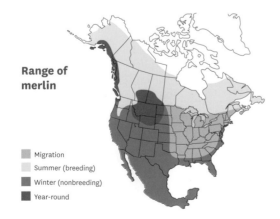

Range of merlin

Migration
Summer (breeding)
Winter (nonbreeding)
Year-round

to cause confusion, although they are slightly smaller and more slender. When the kestrel glides it appears to float, while the heavier merlin tends to sink. Up close, the kestrel can be identified by its distinctive facial stripes. The male is blue-gray with a reddish tail; the female is reddish with black bars.

American kestrels eat mostly large insects, with a preference for grasshoppers, but they also enjoy mice, voles, and other small rodents. About 10 percent of their diet consists of birds, up to about the size of a quail, so they are capable of taking young or small poultry. However, because they are much more keen on bugs and rodents, some poultry keepers put up nest boxes to encourage kestrels to stick around.

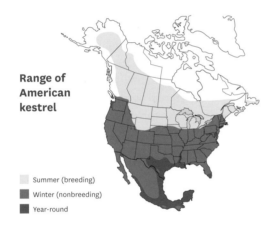

Range of American kestrel

- Summer (breeding)
- Winter (nonbreeding)
- Year-round

Comparing Falcon Shape and Size

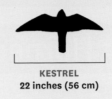

KESTREL
22 inches (56 cm)

MERLIN
24 inches (60 cm)

PRAIRIE FALCON
39 inches (101 cm)

PEREGRINE FALCON
42 inches (105 cm)

GYRFALCON
48 inches (122 cm)

HAWKS AND EAGLES

151

OWLS

Great Horned Owl · Snowy Owl · Great Gray Owl · Barred Owl

Great Horned Owl

AVERAGE LENGTH: 22 inches (55.5 cm)
AVERAGE WEIGHT: 3¼ pounds (1.5 kg)
AVERAGE WINGSPAN: 4 feet (1.2 m)

The most common raptors throughout North America, great horned owls (*Bubo virginianus*) enjoy a more varied diet than any other bird of prey. The largest of our North American owls, they are slightly bigger than red-tailed hawks and are considered the red-tails' nighttime counterparts. They are most active from about dusk to midnight, and then again in the early hours before dawn, although they may hunt in the daytime, particularly on overcast days.

Their widely diverse diet ranges from small rodents to birds the size of a goose. Delectables they feed on include rats, ground squirrels, raccoons, skunks, opossums, hawks, barred owls, and even other great horned owls. They are perch-and-pounce hunters that mostly capture

prey on the ground but prefer, whenever possible, to carry it elsewhere to eat.

Because they hunt mostly in the dark, they rely on their outstanding sense of hearing to locate prey. One ear is slightly higher than the other, causing sound waves to reach the ears at slightly different times. The resulting stereo effect helps an owl pinpoint the location of prey. Further, flat circles of feathers surrounding the eyes function like satellite dishes to catch sound and funnel it toward the ears.

The owl's eyes are tubular in shape and housed in large, round bony structures (called sclerotic rings). Because the eyes aren't round, like a human's, and because they are surrounded by rigid bone, the owl is unable to look around by moving its eyes. Instead, it has to turn its whole head. An extremely flexible neck allows the owl to rotate its head 270 degrees in either direction from forward, a feat that leaves the mistaken impression that an owl can turn its head all the way around.

Unlike most birds, whose wings make a whooshing sound when they fly, an owl flies almost silently. The flight feathers are spaced apart to allow air to freely flow past them, the front, or leading, edges of the primary wing feathers are fitted with comb-like serrations, and the back, or trailing, edge of each primary is softly fringed. Instead of turbulence created by air rushing noisily over the wing, as occurs with other birds, the fringes break it into mini-turbulences, effectively muffling the sound of flight. Since owls hunt by sound as well as by sight, silent flight has two benefits: the owl can hear better to locate prey, and its next meal doesn't hear the owl coming.

Because great horned owls fly silently and are active mostly after dark, you are less likely to see them than to hear their hooting calls at dusk and dawn. Although they hoot year-round, when a pair starts nesting they fill the forest with deeply pitched hoots, four or five in series: *hoo-h'HOO-hoo-hoo*. Pairs often hoot in duet, with the male's hoot slightly deeper in pitch than the female's. Although great horned owls make a variety of other sounds, thanks to their territorial hooting they are known as hoot owls, a distinction they share with barred owls. If you're watching a movie and you hear an owl hoot, however, it's most likely a great horned owl.

TYPICAL SIGNS. Birds make up only about 10 percent of the typical great horned owl's diet, which is no consolation for the poultry keeper whose birds are on the menu. Although mature great horns are rarely interested in poultry, a hungry teenager is a different story. In the fall, when parents tire of feeding their offspring, a young owl may take advantage of chickens or ducks sleeping or nesting in the open, and may even learn to raid an unsecured coop. In either case, the owl most likely will be back for more.

Great horned owls are more at ease walking on the ground than other owl species and therefore are more likely to leave tracks. Like other owls, their claws are oriented with two toes pointing forward and two pointing back, an arrangement known as K-shaped. One of the back toes is reversible, however, and when the owl needs to get a better grasp it turns this toe forward, creating a deadly grip capable of crushing the skull or breaking the neck of prey. It then uses its hooked beak to rip meat off the bones, starting with the head and neck and working down the breast. It may crush and consume smaller bones but usually discards the wings and feet.

If the backyard bird is too heavy to carry, or the owl isn't all that hungry, it may eat just the nutritious brain, leaving the carcass minus only the head. Sometimes an owl will eat part of the breast until the weight has been reduced enough to carry the chicken elsewhere, in which case all you'll find of the missing bird may be a few clumps of feathers. Baby birds, on

---

CLUES: Great Horned Owl

Range of great horned owl

■ Year-round
■ Year-round (scarce)

Pellet

A great horned owl typically expels one tightly compact dark gray or black pellet per day, averaging 1 inch in diameter and up to 4½ inches long (2.5 × 11.4 cm).

Tracks

Large K-shaped bird tracks are most likely those of a great horned owl, which walks more than any other owl species. The tracks may be as much as 4½ inches long and 3 inches wide (11.4 × 7.6 cm), with a stride ranging from 3 to 11 inches (7.6 to 28 cm).

the other hand, can be swallowed whole, leaving little or no evidence.

SEASONALITY. Great horned owls nest earlier than most other raptors. The females may start laying in January — a little later in the far north and earlier in the southern states, depending on the weather and the abundance of available food. Because they typically start laying eggs before trees leaf out in spring, their nesting activities can be fairly easy to spot.

A pair doesn't build its own nest but takes over a nest built by other birds or by squirrels, or they may nest in a cave or tree hollow. While owlets are in the nest, the female guards the brood while the male hunts. By about the middle of May, the nestlings have learned to fly but may still be fed by their parents into autumn. Once they are on their own, the young ones become aggressive about finding things to eat, and that's the time of greatest danger to backyard poultry.

APPEARANCE. A distinguishing feature of mature great horned owls is their horns, which aren't really horns but tufts of feathers that look like horns. Another distinguishing feature is the intimidating stare of their enormous yellow eyes.

Great horns easily maneuver through a forest thanks to their relatively short, wide wings, which may have a span of up to 5 feet. Their legs are feathered down to the talons on their large, powerful feet. The leg covering helps keep them warm in colder climates.

The mottled plumage varies in color from one area to another, ranging from a reddish-brown to a dark gray-brown. The face is reddish-brown and is accented by a tidy white bib.

HABITAT AND RANGE. Great horned owls aren't fussy in selecting a territory as long as it has some trees and some open habitat for hunting. They may settle on farmland, in forests, in the desert, on subarctic tundra, and even in an urban area, such as a city park that features groves of trees.

Northernmost great horns may migrate, but most pairs are year-round residents in their home territory. Being entirely familiar with trees and other features within their territory is one of the ways great horned owls are able to get around so well in the dark.

BEST DETERRENTS. The most effective ways to prevent predation by great horned owls are to close up your poultry at night and keep them in a covered run during the day. The cover must be tight and sturdy. Hardware cloth is ideal, but expensive. Less-expensive tight mesh netting works best in a double layer about 6 inches apart. Strategically placed flashing LED lights are also helpful.

More Owls

Owls are members of the order Strigiformes, a word derived from the Latin *strig* (plural of *strix*), meaning owls, and *formes*, meaning forms — in other words, they look like owls. Of the 250 species worldwide, 19 reside in the United States and Canada. Most species prefer to eat rodents, attack birds in flight, or are too small to bother poultry. Barn owls, for example, typically take up residence in farm buildings, where they pay the rent by clearing out mice and other rodents.

Features all strigiformes share in common are large, round heads with large, forward-facing eyes surrounded by feathered facial disks; wide wings; short tails; feathered legs; an upright posture; and silent flight. They have superior hearing and vision designed for hunting in low light. They cannot move their eyeballs but make up for that by having twice as many neck vertebrae as humans, allowing them to turn their heads to see what's directly behind them. Whereas most birds have three toes pointing forward and one pointing back, all owls are yoke-toed like the great horned owl.

Owls are solitary birds that hunt primarily at night and dine mainly on rodents and small mammals. They have strong feet with sharp talons for grasping, and hooked beaks that can easily snap the neck of prey and rip into flesh. Before the next meal, they regurgitate undigestible portions of the previous meal as a pellet. Their droppings consist mostly of uric acid, splattered like white wash. Southern old-timers are fond of referring to any slippery surface as being "slicker than owl shit."

Most owls have fluffy feathers that are either gray or brown with a mottled or streaked appearance that blends well with tree bark, providing effective camouflage while they sleep during the daytime. Female and male owls of the same species look alike, although the females tend to be slightly larger than the males. Compared to hawks and eagles, owls have a less streamlined build that is designed for short, powerful bursts of flight rather than sustained soaring. As a result, owls don't migrate. But a large owl can carry away a backyard bird weighing up to three times its own weight, while a hawk can lift only about half its body weight.

Of the four species that are large enough to prey on poultry, two are of the genus *Bubo* (horned owls) and two are of the genus *Strix* (wood owls with round faces and no feather tufts). Aside from the great horned owl, they are described here starting with the largest (by weight). Finding a stray owl feather at the site of a kill can help you identify the species. Deterrents and signs are the same as for great horned owls.

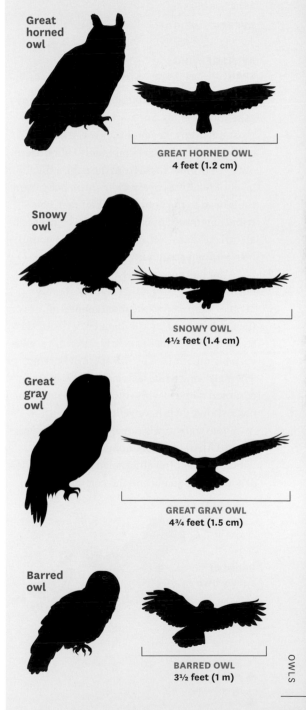

Comparing Owl Shape and Size

Great horned owl

GREAT HORNED OWL
4 feet (1.2 cm)

Snowy owl

SNOWY OWL
4½ feet (1.4 cm)

Great gray owl

GREAT GRAY OWL
4¾ feet (1.5 cm)

Barred owl

BARRED OWL
3½ feet (1 m)

Snowy Owl

AVERAGE LENGTH:
24 inches (60 cm)

AVERAGE WEIGHT:
4 pounds (1.8 kg)

AVERAGE WING-SPAN: 4½ feet (1.4 m)

Snowy owls (*Bubo scandiacus*) are unlike any other owl species. For starters, they are the most northerly. Because they breed on the arctic tundra, they are commonly called arctic owls. To protect them from the cold, a thick layer of feathers covers their entire body from their head to the bottoms of their feet, contributing toward making them the heaviest of all owl species — on average they are 1 pound heavier than great horned owls.

Although snowy owls are classified as horned owls, their feather tufts are barely visible compared to the monumental affairs of the great horned owl. To blend in with arctic snow, their plumage is nearly pure white, with variable amounts of brown spots or barring. Their striking whiteness gives them another common name — white owls. Like great horned owls, snowy owls have yellow eyes.

Some snowy owls remain in the Arctic year-round. Others are more nomadic during the winter months, traveling down into the prairies of southern Canada and the northern United States. Sporadically they engage in so-called irruptions, the owl version of migration, during which they may appear, sometimes in large numbers, anywhere in the lower 48 states — commonly a result of periodic shortages of their dietary staple, lemmings and voles.

During an irruption, they settle in treeless, rolling terrain that reminds them of home. They prefer open fields and farmland where rodents are numerous. When rodents are scarce, they look for other easy prey in abundance. They are powerful birds, capable of taking poultry as large as geese. Still, if I ever see a snowy owl I will feel greatly privileged.

Great Gray Owl

AVERAGE LENGTH:
28½ inches (72 cm)

AVERAGE WEIGHT:
2½ pounds (1.1 kg)

AVERAGE WING-SPAN: 4¾ feet (1.5 m)

Although great gray owls (*Strix nebulosa*) are capable of killing sharp-shinned hawks, they rarely prey on birds of any kind but much prefer voles and other small mammals, including weasels. For a chicken keeper, making a choice between great gray owls and weasels should

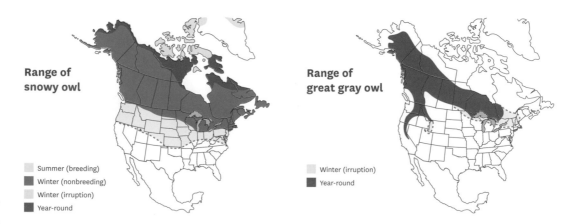

Range of snowy owl

Summer (breeding)
Winter (nonbreeding)
Winter (irruption)
Year-round

Range of great gray owl

Winter (irruption)
Year-round

be a no-brainer. If I were lucky enough to live within this owl's home range, I'd try to attract a breeding pair by constructing a nest platform according to the great gray owl nest structure plan available at *www.nestwatch.org*.

Barred Owl

AVERAGE LENGTH: 19 inches (48 cm)
AVERAGE WEIGHT: 1⅔ pounds (0.75 kg)
AVERAGE WINGSPAN: 3½ feet (1 m)

After great horned owls, barred owls (*Strix varia*) are the second most problematic owl predators of poultry, but they are not nearly as widespread. They are slightly smaller than great horned owls, and less aggressive. They are concentrated primarily in the eastern half of the United States and in the Pacific Northwest, where they establish territory in large tracts of mature and old-growth forests. Where their range overlaps with that of great horned owls, the most serious predatory threat to the barred owl is the great horned owl.

The gray-brown barred owl is easily distinguished from the great horned by its round, tuftless head, soulful brown eyes, horizontally barred bib, and vertically striped chest. In the eastern states it is unlike any other owl species, but in the West it looks much like its less numerous cousin, the spotted owl (which rarely, if ever, preys on poultry).

Although barred owls do most of their hunting at sunset and into the night, they also often hunt during the day. Like great horned owls, they enjoy a varied diet that includes small mammals, amphibians, reptiles, and birds. They are likely to eat smaller poultry on the spot and carry away larger birds to consume from a perch, eating the head first, then the breast, and then— if still hungry — working on the rest.

Range of barred owl

■ Year-round

NIGHT MUSIC

Like great horned owls, barred owls are sometimes referred to as hoot owls. Their presence in your area easily may be detected by listening for their distinctive eight-note hoot, which bird watchers are fond of pointing out sounds like "Who cooks for you? Who cooks for you all?" If you learn to imitate the sound, you might be rewarded by a territorial owl coming to investigate. The barred owl's repertoire of additional sounds is more varied than that of any other North American owl and includes barks, cackles, grumbles, hisses, shrieks, squeaks, twitters, and loud beak snapping. When a pair courts, usually in February or March, their boisterous duet fills the forest with eerie sounds.

Things That Go "Thump" in the Night

In the middle of night, my husband Allan and I were awakened by a single loud thump on the roof over our bedroom. Next morning, one of our young guinea fowl was missing.

We had a flock of mostly pearl guineas, with a few white ones in the lot. White poultry of any kind are usually the first to succumb to predation, and our white guinea fowl were no exception — the missing bird was a white one.

These young guineas were supposed to spend nights safely inside the coop with the chickens. Instead, they preferred to sleep in the open, perching along the utility line that runs from a power pole near the garden to the corner of the roof above our bedroom.

A few nights later we heard another thump. I jumped out of bed and, in the bright moonlight, saw what looked like a stout young boy in baggy pants standing in the backyard. After a moment the "boy" spread his wings and turned into a great horned owl that flew away. The thumping sound was its wing hitting the roof.

In the morning, we discovered that one of our white guineas had been mortally wounded. The owl had apparently grabbed the guinea by the neck, partially severing its windpipe, but hadn't been able to keep a grip on the guinea as it attempted to fly away with its prize.

Beneath the utility line where the guineas roosted was a woodshed. To entice the birds to roost under the shed's roof, we installed a perching pole that stuck out a few feet into the open. At dusk, as the guineas were getting settled on the utility line, Allan went out and talked quietly to them, asking them to please come down and sleep in the woodshed. To our surprise, one by one the guineas flew down from the utility line onto the woodshed roof, and from there onto the extended pole.

The next evening they were back up on the utility line. Allan went out and said sternly, "Get down!" To our utter amazement, they did.

Every couple of days, we cut a small piece off the end of the pole, until eventually the entire perch was under the roof, and all the guineas were roosting on it. Although the great horned owl maintained its nightly rounds, it didn't get any more of those guinea fowl.

RACCOON FAMILY

Raccoon · Coati and Ringtail

Raccoon

AVERAGE LENGTH (including tail): 33 inches (84 cm)
AVERAGE HEIGHT AT SHOULDER: 11 inches (28 cm)
AVERAGE WEIGHT: 15 pounds (6.8 kg)

Raccoons (*Procyon lotor*) are considered to be the mammal equivalent of great horned owls. Both are at once adorable and the bane of poultry keepers. Both may be found throughout the United States and into Canada. Both are adaptable in their choice of habitat and are largely nocturnal but may be active at any time of day.

Like great horned owls, raccoons enjoy a widely diverse menu. In nature, a large percentage of their diet consists of berries, nuts, and seeds, yet the chewing surface of their teeth is not as wide as that of a typical herbivore. Although raccoons also eat meat — including fish, frogs, and rodents — they lack the sharply pointed teeth of a carnivore. In rural areas, 'coons typically dine on corn, sorghum, and other cultivated crops. In urban areas, they survive by dumpster diving. They relish pet food, including chicken feed. They love chicken or duck eggs and will also munch on poultry.

A raccoon's front and back paws have five digits and leave prints looking like those of a human baby's hands and feet. The word *raccoon* comes from the Algonquian word *arahkunem,* meaning "he who scratches with his hands." The scientific name *lotor* is Latin for laundryman (no one is sure of the reasoning behind the designation *Procyon,* which is the brightest star in the constellation Canis Minoris).

A common belief is that raccoons use their dexterous hands to wash their food before eating it, which isn't entirely true. In the wild, raccoons often use their hands to search among underwater rocks for fish, frogs, crayfish, mollusks, and other tasty tidbits. Their tactile sense is so well developed that, through feel alone, they are able to identify unseen food items. In captivity, 'coons tend to satisfy the instinct to forage in water by dunking and rolling their food, toys, and other objects in water. To enhance tactile sensation, they may rub dry food between their hands and even rub their empty hands together.

Their extreme manual dexterity often gets raccoons into trouble with poultry keepers. Thanks to their long nimble fingers, they are able to untie knots, unlatch gates, and open windows and doors. They are also adept at swimming, climbing, and digging. Raccoons are thus able to gain entry to poultry yards and chicken coops protected by security measures that would defy most other predators.

TYPICAL SIGNS. Raccoons are often attracted to poultry yards — not to the birds themselves, but to feed, especially if it's left out overnight. When I raised chicks in an elevated outdoor brooder, and they persisted in spilling starter ration from their feeder, a pair of young raccoons regularly foraged underneath. Although we frequently saw the 'coons cleaning up the feed, they never bothered the chicks. It's a good thing, too, because a 'coon can reach through wire to grab a chick, pulling out and eating whatever part it gets hold of.

Raccoons inside a coop or run can wreak havoc. Besides relishing eggs, they may also target poultry. They are extremely wasteful in eating just the crop of a chicken or duck,

maybe a little bit of the breast, possibly the entrails, and leaving the better part of the meat. The crop contains mostly grains and invertebrates, which make up the majority of the average raccoon's diet. Since the 'coon's teeth are not designed for tearing meat, after consuming the crop of one chicken or duck, a still-hungry raccoon may leave the rest and go for another. A greater amount of meat eaten from the bone may indicate a dearth of other food resources or may be the work of more than one 'coon.

Although raccoons typically hunt alone, a mother with growing kits may work as a family team. Like weasels and dogs, they may kill multiple birds. If they can't get to chickens or

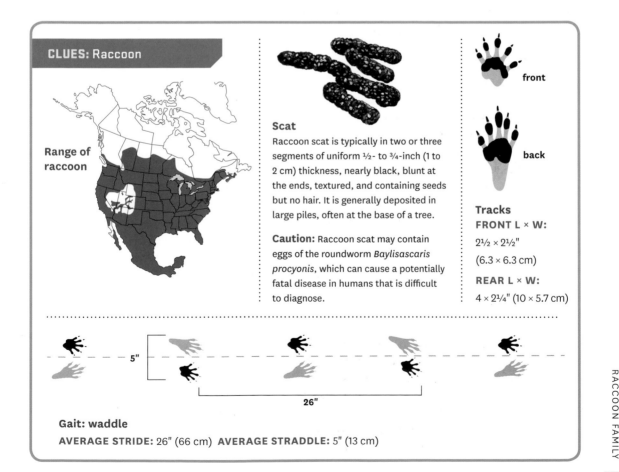

CLUES: Raccoon

Range of raccoon

Scat
Raccoon scat is typically in two or three segments of uniform ½- to ¾-inch (1 to 2 cm) thickness, nearly black, blunt at the ends, textured, and containing seeds but no hair. It is generally deposited in large piles, often at the base of a tree.

Caution: Raccoon scat may contain eggs of the roundworm *Baylisascaris procyonis*, which can cause a potentially fatal disease in humans that is difficult to diagnose.

front

back

Tracks
FRONT L × W:
2½ × 2½"
(6.3 × 6.3 cm)

REAR L × W:
4 × 2¼" (10 × 5.7 cm)

5"

26"

Gait: waddle
AVERAGE STRIDE: 26" (66 cm) **AVERAGE STRADDLE:** 5" (13 cm)

ducks protected inside a fence, they will try to scare the birds into sticking their heads through the fence and then bite off the birds' heads. But headless chickens are not necessarily a sign of raccoon predation. Other animals enjoy that nutritious delicacy — the brain — loaded with protein, fat, and other nutrients. Headless chickens could be the work of an owl or hawk, or maybe a skunk or a weasel, indicating the need to sleuth for more clues.

Identifying raccoons by listening can be tricky because they make dozens of different sounds, including purring, chittering, growling, snarling, and hissing. Baby raccoons mew, cry, and whine. 'Coons calling to each other sound like whistling screech owls (which do not prey on poultry).

APPEARANCE. The raccoon is the masked bandit of the predator world. Along with its black-masked face, a distinguishing feature is its bushy tail, which is about half the animal's total length and sports four to seven alternating black and white rings. The salt-and-pepper body fur ranges in color from light gray to nearly black.

The back legs are longer than the fronts, giving the 'coon an unmistakable hunched look as it lumbers along. The back feet can rotate 180 degrees, allowing the raccoon to climb a tree or fence upward or downward facing either backward or forward.

STATE ANIMAL

The raccoon is the official state wild animal in Tennessee and the official state furbearer animal in Oklahoma. The ringtail (*Bassariscus astutus*) is the official Arizona state mammal.

Adults range in length from 26 to 39 inches, including the tail, in shoulder height from 9 to 12 inches, and in weight from 6 to 25 pounds. Males are larger than females, and northern populations are larger than those in the south. The largest wild raccoon on record weighed 62.6 pounds and had a total length of 55 inches.

HABITAT AND RANGE. Although raccoons are highly adaptable, they prefer to live near open water with trees, underbrush, or other available cover. They will move into barns, attics, abandoned buildings, and other structures. They are, in fact, becoming more numerous in urban and suburban areas where food and other resources are readily available. So many raccoons live in Toronto, for example, that it has been called Raccoon City and the Raccoon Capital of the World.

City raccoons can be more problematic than their country cousins because they are less afraid of humans and therefore more aggressive. A confronted raccoon typically runs for cover, which is a good thing because raccoons can carry more than a dozen pathogens that are harmful to humans. But when cornered, a raccoon will fight ferociously to protect itself.

BEST DETERRENTS. To minimize raccoon predation, surround the coop and run with electric fence, or cover the run with hardware cloth. Secure doors and gates with two-way latches. Install apron fence to discourage digging. To keep 'coons from getting over a fence by climbing a nearby pole or tree, circle it with 16-inch-wide metal roof flashing, at least 3 feet above the ground. At night, confine poultry inside a coop with all ventilation openings secured by hardware cloth. A dog in the poultry yard overnight will deter raccoons.

A Perpetrator with Accomplices

Bethany Caskey, Iowa, illustrator, *The Chicken Encyclopedia* by Gail Damerow and other books

When we moved to a new farm, we installed a brand-new chicken house and a chicken yard for the flock to enjoy under shade trees. The new chicken fence was 2 × 4-inch welded wire, 5 feet high, with a top and bottom rail and a matching walk-through gate. I felt confident the chickens were securely protected.

But just three days before one of our setting Iowa Blue hens was due to hatch her eggs, something robbed the nest inside the coop within the fenced area. My fault for forgetting to shut the pop door at dusk. All the eggs were gone and the setting hen was dead with her belly opened.

I got a live cage-style trap and, after shutting the chickens in for the night, set the trap outside of the pop door baited with a tin of cat food. By morning, one of the perpetrators was securely inside the trap. Three other small accomplices sat in a low crotch of a nearby tree waiting for mama to take them home.

My plan had been to catch the chicken-killing egg thief and, with the local game warden's permission, transport it several miles to a state forest for release. Now, with three unweaned babies, I was stumped. The kits crawled down the tree to get a better look at me and be closer to Mom. They were cautious, though not afraid, but I knew better than to try to wrestle them into a cage for transportation.

I opened the trap door and mama raccoon scooted out. The kits scrambled down the tree to her side and, in an organized group formation, they scaled the fence and out to freedom. They had illustrated plainly how ineffective a mere 5-foot fence is for an expert climber.

To prevent recurrence, I added an alarm to my phone that goes off at dusk to remind me to shut the pop door each night. I also ran an electric wire on stand-offs attached vertically to each fence post around the enclosure and hooked it to a solar fence charger by the entrance gate. So far, we have had no more losses from raccoons or other predators.

Coati and Ringtail

Coati

COATI
AVERAGE LENGTH (including tail): 4 feet (120 cm)
AVERAGE HEIGHT AT SHOULDER: 10 inches (25 cm)
AVERAGE WEIGHT: 15 pounds (7 kg)

RINGTAIL
AVERAGE LENGTH (including tail): 2½ feet (76 cm)
AVERAGE HEIGHT AT SHOULDER: 6 inches (16 cm)
AVERAGE WEIGHT: 2½ pounds (1.15 kg)

The raccoon family (Procyonidae) originated in the American tropics. Of the 18 species, only raccoons have spread as far north as southern Canada. Two others, coatis and ringtails, make their home in limited areas of the United States. Relics of the procyonids' tropical origins are their long fingers, their furless soles, their fondness of fruits and berries, and their tendency to climb trees to seek food and safety.

Other features that raccoon family members share include:

Range of coati

- small to medium size
- brown or gray body fur
- distinctive facial markings
- banded tail the same length as the body
- five digits on all four feet
- short, curved claws
- ankles that can rotate 180 degrees
- sole walkers (plantigrade)
- males do not help care for the young

Ringtail

Compared to raccoons, which vary in their degree of sociability, ringtails are mostly solitary, while coatis are highly gregarious. Ringtails, like raccoons, are nocturnal; coatis are diurnal. Deter coatis and ringtails by the same methods as for raccoons.

Range of ringtail

RACCOON FAMILY

DOG FAMILY

Red Fox · Gray Fox · Kit Fox · Swift Fox · Arctic Fox · Coyotes · Gray Wolf · Domestic Dog

Red Fox

AVERAGE LENGTH: male, 3 feet (1 m);
female, 2 feet (60 cm)
AVERAGE HEIGHT: 17 inches (43 cm)
AVERAGE WEIGHT: male, 14 pounds (6 kg);
female, 11 pounds (5 kg)

Notoriously, the red fox (*Vulpes vulpes*) has a penchant for poultry. More widely distributed than any other carnivore, it is at home throughout Alaska, Canada, and most of the lower 48, except in parts of the West and Southwest. Although five other fox species live in North America, when people speak of a fox they typically mean a red fox.

Red foxes are doglike in appearance but catlike in character. Like a dog, the fox has a longish pointed muzzle, long pointed ears, and claws that are not fully retractable. Like a cat, a fox is curious and playful and typically hunts alone. It stalks prey and pounces like a cat, and it may play with dinner before eating. Like a cat (and unlike a dog), the fox won't overeat but prefers several small meals throughout the day.

Like a house cat, the fox has vertical pupils rather than the round pupils of other members of the dog family, a feature that improves the visibility of movement and color of prey in low light and helps it gauge distance for pouncing. Like a cat, the fox pins down prey with its front paws and has sensitive whiskers on the backs of its front legs through which it can detect the prey's movements. Unfortunately for cats, felines are among the many items on the fox's menu.

Depending on seasonal availability, the omnivorous red fox eats grains, nuts, fruits, berries, mushrooms, insects, worms, amphibians, reptiles, and carrion. But its dietary preference is for rodents, rabbits, and birds, including poultry and eggs. When feeding young, the fox buries surplus food near its den.

TYPICAL SIGNS. The red fox is partial to chickens, ducks, guinea fowl, and pheasants, usually taking only one at a time. More than one bird missing could be the work of a pair, perhaps with offspring. A fox typically snatches a bird by the throat and trots off with it. After the initial surprise, the captured bird normally won't struggle, so the only remaining evidence might be a few feathers and maybe a few drops of blood.

When the fox finds a safe place to dine, it first eats the breast and legs. As it strips meat from the leg bones, pulled tendons leave the bird's toes curled. Uneaten parts may be either scattered or buried. Sometimes all that's left are the wings.

Sometimes multiple birds will be killed in a single frenzied coop raid (see "Killing Sprees" on page 36). Headless carcasses may be scattered around, perhaps partially buried in loose litter or soil. Or all that's left might be piles feathers.

An unprotected setting hen is especially vulnerable, as a fox will target both the hen and her eggs. After the fox sets aside the hen's body, it will open egg shells just enough to lick out the contents, leaving empty shells near the nest before carrying the hen away to enjoy elsewhere.

The oval tracks left by a red fox are similar to those of a small dog. Identifying features include four visible toes, sometimes visible claw marks, hair between the toes, open space between the toes and heel pad, and a V-shaped bar across the heel pad (which no other member of the dog family has). Fox tracks, like those of a cat, commonly appear in a straight line, often with the rear track on top of the front one (direct register).

A fox tends to be active after dark but, like a cat, keeps flexible hours if it feels secure.

Mature foxes typically hunt alone. Teenagers may hunt in pairs. Occasionally foxes hunt in family packs. Family members stay in contact by calling each other, sounds you may hear at night. Red foxes have many different sounds that include barking, screaming, and whining. Some calls are owl-like; others are coyote-like.

SEASONALITY. Foxes breed in winter and kits are born in early to midspring. The female (vixen) stays in the den, protecting the offspring until they are weaned, while the male (dog fox) provides food. When a male breeds more than one female, the vixens may den together and raise their kits together, keeping the dog fox busy bringing home groceries. The more kits he has to provide for, the greedier he is for poultry.

Each vixen gives birth to four or more offspring, varying with the availability of resources. When the kits are about a month old,

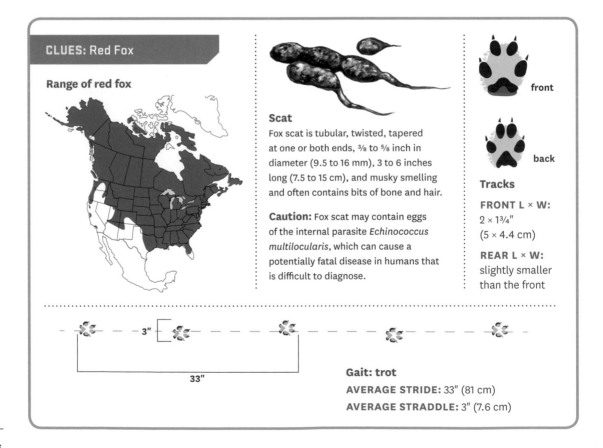

CLUES: Red Fox

Range of red fox

Scat

Fox scat is tubular, twisted, tapered at one or both ends, ⅜ to ⅝ inch in diameter (9.5 to 16 mm), 3 to 6 inches long (7.5 to 15 cm), and musky smelling and often contains bits of bone and hair.

Caution: Fox scat may contain eggs of the internal parasite *Echinococcus multilocularis*, which can cause a potentially fatal disease in humans that is difficult to diagnose.

front

back

Tracks

FRONT L × W:
2 × 1¾"
(5 × 4.4 cm)

REAR L × W:
slightly smaller than the front

3"

33"

Gait: trot
AVERAGE STRIDE: 33" (81 cm)
AVERAGE STRADDLE: 3" (7.6 cm)

the mother begins feeding them bits of meat. Soon she'll bring mice and other small live animals and insects for them to practice killing. When the kits are about two months old, the vixen shows them how to hunt.

By fall, the kits are six months old and ready to hunt on their own. This season is the second-most dangerous time for fox predation, because the young aren't yet cautious about hunting during the day.

APPEARANCE. The quintessential red fox has rusty-red to orange fur, a white underbelly, and four black boots. But red foxes come in other colors, including black, silver (black with white hairs), and cross (rusty red with a dark stripe along the back and a dark band across the shoulders, forming a cross). Red foxes of differing colors may be born in the same litter. An identifying feature, regardless of coat color, is the white-tipped tail.

The fox has short legs, a proportionately long body, and a bushy tail that makes up about one-third of its total length. The tail keeps the nose and feet warm during sleep and is used for balance when the fox runs. The red fox can run as fast as 30 miles an hour and, when confronted, would rather flee than fight.

The red fox is the largest of the fox species; males are larger than females and northern populations are somewhat larger than those in warmer areas. Mature males are about 3 feet long, nose to tail, while females are about 2 feet long. Weights vary from about 6½ to 24 pounds, with an average of about 14 pounds for males and 11 pounds for females.

HABITAT AND RANGE. Red foxes, being highly flexible in their diet, adapt readily to both urban and rural environments. They generally prefer woodlands with some open areas for rodent hunting, but in cities they take to dumpster diving and dining on roadkill.

The size of a fox's home range depends on the availability of resources — food, water, and safe den sites for raising young. Generally each territory has more than one den, and previous dens may or may not be reused. The fox rarely digs its own den but prefers to remodel a badger or groundhog burrow. Between litters, the rural fox generally sleeps on the ground in the open; the urban fox curls up under a protective outbuilding or other structure.

BEST DETERRENTS. Keeping foxes out of your henhouse presents a challenge. They can squeeze through openings as small as 3 inches. They can leap over a 6-foot-high fence. They are excellent diggers and good swimmers.

An electric fence with a tight-fitting gate is the best deterrent. Next best would be extension arms at the top of the fence and a buried barrier at the bottom of the fence. Constantly patrol any fence to watch for breaches. Lock up poultry as soon as they go to roost and don't let them out until mid- to late morning.

Foxes love rodents, and rodents love chicken feed. Removing feeders from the yard at night will make it less attractive to rodents and therefore to foxes. A guardian dog is helpful, since foxes steer clear of potential danger to avoid injury leading to an inability to care for young. Removing a problem fox is not a reasonable option — another fox would soon take over the vacant territory.

STATE ANIMAL

The red fox (*Vulpes vulpes*) is one of Mississippi's official state mammals. The grey [sic] fox (*Urocyon cinereoargenteus*) is the official Delaware state wildlife animal, selected at the urging of fourth-grade students on the basis that gray foxes do not hibernate and so are "always ready like our soldiers at Dover Air Force Base."

Diurnal Foxes

Shortly after my husband and I moved to our Tennessee farm, I had my first experience with red foxes. We were working in the garden and had let our New Hampshires out to forage. Instead of staying near the garden, they wandered down the wooded slope toward the creek. Soon we heard a startled squawk, shortly followed by another startled squawk. I said, "A pair of foxes just snatched two of our hens." Sure enough, when the flock came back up the hill, two were missing. Thus ended their free-range privileges until we could provide a yard with an electric fence.

Outside our kitchen door is a pond on which we decided to keep a few mallard ducks. One day, I looked out the window to see a red fox eyeing the ducks in the pond. As the fox approached, the ducks grouped tightly together and swam around the pond's edge. The fox followed on the shore until the ducks had made two complete circuits around the pond, whereupon the fox changed its mind about having duck for lunch and disappeared into the woods. Although foxes are excellent swimmers, this one apparently wasn't willing to take on a flotilla of ducks.

Another time it was our guinea fowl that got the best of a fox. A gang of guineas was wandering down the driveway when a fox came out of the forest thinking to grab an easy meal. When the fox lunged at one guinea, other guineas rushed at it from behind. The fox whirled around to confront the attackers and found itself encircled by guinea fowl. Each time the fox lunged at one guinea, it was attacked from behind by others. This scenario went on for quite some time until the fox decided the better part of valor was to tuck his tail between his legs and slink away.

Over the years, we've occasionally had poultry disappear in the wee hours of morning, often no doubt thanks to our resident foxes. However, all three of the above incidents involved red foxes that apparently feel perfectly safe wandering our farm in broad daylight.

More Foxes

Foxes are the smallest members of the dog family, Canidae, which in North America includes coyotes and wolves, as well as domestic dogs. As a group, canids typically have long muzzles, with jaws and teeth designed for tearing flesh and breaking bones. Although they are primarily meat eaters, they tend to be more omnivorous than other carnivores.

Foxes have upright ears and acute hearing, as well as an acute sense of smell, although their vision is not as keen as that of most carnivores. They have muscular bodies and long legs for endurance when chasing prey. Canids have five toes in front and four in back, and they walk on their toes. Their blunt claws are not retractable, like a cat's, but are used for traction rather than for fighting or killing prey. They tackle prey by the back of the neck. While most canids shake the victim to demobilize it, foxes, like cats, dispatch prey by biting.

North American foxes are classified into two genera, *Vulpes* (with four species) and *Urocyon* (two species). The most common *Vulpes* is the red fox; the most common *Urocyon* is the gray fox. Among the six species that inhabit the United States and Canada, red and gray foxes are more likely than others to prey on poultry. Deterrents for all foxes are mostly the same as for red foxes, with the exception that gray foxes must be discouraged from climbing.

Comparing Fox Shape and Size

Red fox

Gray fox

Kit fox

Swift fox

Arctic fox

Gray fox

Kit fox

Gray Fox

AVERAGE LENGTH (including tail): 3¼ feet (1 m)
AVERAGE HEIGHT AT SHOULDER: 14 inches (35 cm)
AVERAGE WEIGHT: 12 pounds (5.5 kg)

Sometimes called the tree fox, the gray fox (*Urocyon cinereoargenteus*) is the only canid in North America able to climb trees — which bodes poorly for guinea fowl or chickens that roost in trees at night. The gray fox pretty much enjoys the same varied diet as a red fox, although it generally eats more plants and fish.

Gray foxes are less widespread than red foxes, ranging from Canada's extreme south throughout the United States, except in the Great Plains and the mountains of the Northwest. They are more nocturnal and secretive than red foxes, preferring forest habitats or dense brush offering protective cover, and they maintain more limited territories. Grays are less tolerant of human activity than red foxes and therefore less commonly seen in agricultural or suburban areas.

Kit Fox

AVERAGE LENGTH (including tail):
28 inches (72 cm)

AVERAGE HEIGHT AT SHOULDER:
11 inches (28 cm)

AVERAGE WEIGHT: 4½ pounds (2 kg)

Kit foxes (*Vulpes macrotis*) inhabit desert and semi-arid grasslands from southern California to western Colorado, and from western Texas northward into southern Oregon and Idaho. An alternative name for the kit fox is desert fox. The kit designation for this diminutive species supposedly is short for kitten.

In the summer, the kit fox has a rusty-tan or buff-colored coat to blend in with the dry environment; in winter the coat turns more grayish. The underbelly is pale and the tip of the tail is black. The kit fox has oversize ears to help dissipate body heat and dense hair on the soles to protect the paws from hot desert sand.

Baby kits are typically born in February or March, are weaned in about a month, and begin hunting independently at about six months of age. Kit foxes are primarily carnivorous, preferring ground squirrels and other rodents, rabbits, lizards, snakes, insects, and birds, as well as eating carrion and some fruit. These foxes are not readily intimidated by humans and will take up residence in commercial orchards and sometimes urban habitats. Unlike other foxes, which run when confronted, the kit fox will flatten to the ground.

Swift Fox

AVERAGE LENGTH (including tail):
31 inches (80 cm)

AVERAGE HEIGHT AT SHOULDER:
12 inches (30 cm)

AVERAGE WEIGHT: 5½ pounds (2.5 kg)

Named for its speed, the swift fox (*Vulpes velox*) can run at 35 miles per hour or more, about three times faster than the average human can run (*velox* is Latin for swift). Native to the Great Plains, swift foxes today are found in isolated areas from southwestern Canada to northern Texas — thanks to such things as human development in their grassland habitat, coyote predation, and poison bait intended for coyotes.

Swift fox

Arctic fox

In the northern areas, swift fox kits are born in mid-May; southward, kits are born a little earlier, in March or early April. They are weaned at about six weeks of age and start hunting on their own in the fall. A swift fox's diet varies with seasonal availability of prey, which includes rabbits, prairie dogs, ground squirrels, mice, insects, reptiles, and amphibians. A swift fox also eats berries, fruit, nuts, seeds, grasses, and carrion. And, of course, birds and eggs.

Arctic Fox

AVERAGE LENGTH (including tail):
3 feet (90 m)

AVERAGE HEIGHT AT SHOULDER:
10½ inches (25 cm)

AVERAGE WEIGHT: 10 pounds (4.5 kg)

Inhabiting the arctic tundra from western Alaska east through northern Canada, the arctic fox (*Vulpes lagopus*) is an omnivore and will eat whatever it can find, which is mostly lemmings, supplemented by ground squirrels, insects, berries, carrion, birds, and eggs. Wherever possible, it takes full advantage of the increased prevalence of backyard chickens.

In summer, the arctic fox wears a dark brown coat. To blend in with ice and snow in winter, its warm, thick coat changes to white, giving it two of its common names — white fox and snow fox. It has short legs, short ears, and a short nose, all of which help reduce heat loss in cold temperatures. The thick fur on its paws provides traction on ice and snow, and it wraps its bushy tail around its body to keep itself warm while sleeping.

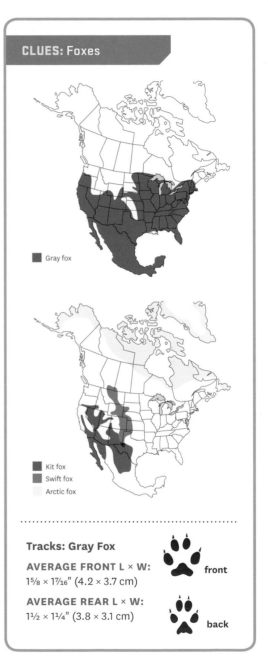

CLUES: Foxes

Gray fox

Kit fox
Swift fox
Arctic fox

Tracks: Gray Fox

AVERAGE FRONT L × W:
1⅝ × 1⁷⁄₁₆" (4.2 × 3.7 cm)

front

AVERAGE REAR L × W:
1½ × 1¼" (3.8 × 3.1 cm)

back

Coyotes

WESTERN COYOTE
AVERAGE LENGTH (including tail): 4 feet (1.2 m)
AVERAGE HEIGHT AT SHOULDER: 18 inches (46 cm)
AVERAGE WEIGHT: 25 pounds (11 kg)

EASTERN COYOTE
AVERAGE LENGTH (including tail): 4½ feet (1.4 m)
AVERAGE HEIGHT AT SHOULDER: 22 inches (55 cm)
AVERAGE WEIGHT: 35 pounds (16 kg)

Coyotes (*Canis latrans*) make their home throughout most of Canada and the United States, including Alaska. They adapt to all habitats, including woodlands, prairies, and big cities. They are considered to be nocturnal but can be active during the day, especially when young. Mature animals feeding pups and those that feel safe around humans also come out during daylight.

STATE ANIMAL

The coyote (*Canis latrans*) is South Dakota's official state animal.

Although coyotes and foxes compete for the same resources, and coyotes prey on foxes, they can and do coexist. On our farm, we see coyotes as well as both red and gray foxes. Able to run at speeds of 43 miles per hour, a coyote can catch a red fox, whose top speed is 30 miles per hour. A gray more easily evades a coyote by climbing a tree.

Coyotes typically hunt alone or in pairs, and sometimes in family groups. They enjoy a widely varied diet and will eat just about anything that's handy, from carrion and garbage to pets and poultry. Their preference is for rabbits and rodents, and they sometimes bury surplus to eat later.

TYPICAL SIGNS. Poultry predation is most likely to occur in spring, when pups are being fed; in summer, when pups are learning to hunt; and in fall, when inexperienced juveniles strike out on their own. Where a coyote has easy access to poultry, it may make off with an entire bird, perhaps leaving a trail of feathers if it nabs a turkey or goose too heavy to carry without dragging. The coyote commonly eats the whole bird, leaving the skeleton mostly intact, just as a raptor would. The difference is

Western coyote

Eastern coyote

that the coyote slides the carcass around while pulling meat off the bone, scattering feathers randomly instead of arranging them in a neat circle.

BEST DETERRENTS. Coyotes can jump over an 8-foot fence, climb up and over a taller fence, dig under a fence, or chew through wire lighter than 18 gauge. However, they tend not to bother poultry unless access is easy or they can find little else to eat. Where coyotes become an issue, an electric fence or a chain-link fence with extension arms or rollers at the top are the best options for protecting poultry. A guard dog is also helpful. Scent deterrents — such as wolf or coyote urine, or ammonia-soaked rags — provide temporary protection, as do motion-activated devices such as flashing lights, noisemakers, and water sprinklers.

If a particular coyote becomes troublesome, removing that animal may be necessary, especially since it may teach others its bad habit. Some states allow coyote hunting, while other states protect them. However, attempting to eliminate coyotes in general doesn't work; they will rapidly repopulate a vacant territory. An emerging science uses recordings of a coyote chorus to fool resident coyotes into thinking they have increased competition for available resources, thus reducing their rate of reproduction.

CLUES: Coyote

Range of coyotes

■ Western coyote
□ Eastern coyote

Scat
Less than 1 inch (2.5 cm) diameter. Contains hair.

front

Tracks, Western Coyote

AVERAGE FRONT:
2¾ × 2" (7 × 5 cm)

AVERAGE REAR:
2½ × 1⅝" (6.4 × 4 cm)

back

Tracks, Eastern Coyote

AVERAGE FRONT:
3 × 2¼" (7.8 × 5.7 cm)

AVERAGE REAR:
2¾ × 2" (7 × 5 cm)

Gait, Western Coyote: trot
AVERAGE STRIDE: 36" (91.4 cm)
AVERAGE STRADDLE: 3⅛" (7.7 cm)

Gait, Eastern Coyote: trot
AVERAGE STRIDE: 42½" (108 cm)
AVERAGE STRADDLE: 3¾" (9.45 cm)

Gray Wolf

AVERAGE LENGTH (including tail): 5½ feet (17 m)
AVERAGE HEIGHT AT SHOULDER: 32 inches (81 cm)
AVERAGE WEIGHT: 80 pounds (36 kg)

The gray wolf (*Canis lupis*), or timber wolf, is by far the most numerous North American wolf species and is the largest wild member of the dog family. The heaviest wolf on record weighed 175 pounds when killed in Alaska in 1939. The color is usually gray or brown, although the arctic subspecies is white, and some gray wolves are black. Gray wolves, in general, are about twice the size of a coyote, but a young gray easily might be mistaken for a coyote. Further, a coyote's thick winter coat makes it look bigger than it is, and where coyote-wolf hybrids occur, the distinction is further blurred.

Gray wolves are found in a wide range of habitats throughout much of Canada but are rare in the United States, except along the western half of the Canadian border. They generally prefer areas away from human development.

When hunting as a pack, grays prey on large animals like moose and reindeer. Solo hunters concentrate on smaller prey, such as rabbits and beavers. Wolves may be attracted to poultry free-ranging near a forest too small to support a sufficient supply of deer prey. Chickens are less of interest than larger turkeys and geese. Poultry attacks are most likely to occur at dawn or dusk, although in winter when snow reflects light, wolves become more active during the night.

BEST DETERRENTS. Deterring wolves, especially those hunting in packs, requires a secure fence, preferably electric. Options include guard dogs, flashing LED lights, and fladry.

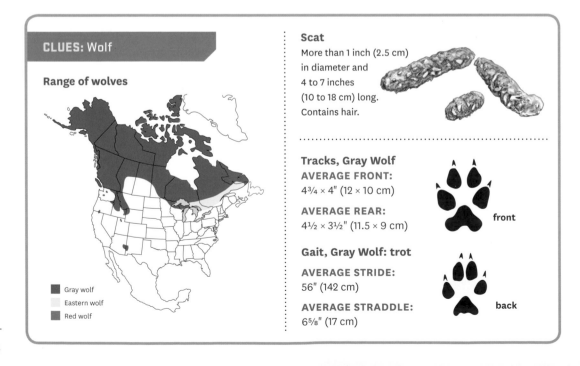

CLUES: Wolf

Range of wolves

- Gray wolf
- Eastern wolf
- Red wolf

Scat
More than 1 inch (2.5 cm) in diameter and 4 to 7 inches (10 to 18 cm) long. Contains hair.

Tracks, Gray Wolf
AVERAGE FRONT:
4¾ × 4" (12 × 10 cm)

AVERAGE REAR:
4½ × 3½" (11.5 × 9 cm)

front

Gait, Gray Wolf: trot

AVERAGE STRIDE:
56" (142 cm)

AVERAGE STRADDLE:
6⅝" (17 cm)

back

Domestic Dog

Domestic dogs (*Canis familiaris*) originated as domesticated wolves and — like chickens, which were domesticated from wild jungle fowl — have been selectively bred into a broad range of sizes, shapes, and colors. Many dog breeds have retained their wolf-like propensity to surplus-kill poultry. Indeed, a fellow chicken breeder once attributed her lucrative repeat business to domestic and feral dogs.

Feral dogs inhabit all 50 states wherever food and escape cover are available. Also called free-ranging dogs, they are domestic dogs, or the descendants of domestic dogs, that have been abandoned or otherwise gone wild, scavenging and hunting like coyotes. Wary of humans, feral dogs are generally active between dusk and dawn. A pack may consist of several breeds and therefore, unlike coyotes, may leave tracks of varying sizes.

TYPICAL SIGNS. Domestic dogs kill for fun, rarely for food, and can be active at any time of day or night. In a poultry yard, they are likely to engage in frenzied killing, tormenting, mutilating, and killing as many birds as they can catch. Skin a dead bird with no obvious sign of injury, and you will likely find red spots where a dog's pointed teeth bruised the skin without breaking through, indicating that the bird was tossed around until it died.

Other signs are a damaged fence or coop and hair clumps on a wire fence. Poultry in elevated cages may have their legs pulled through and bitten off. You can pretty much bet that if you tell a dog's owner that the dog maimed or killed your poultry, the person will vehemently deny it — even if you saw the dog do it and have proof.

Compared to the tracks of a fox or coyote, a dog's tracks are round (rather than oblong), all toenail marks are visible, and the toes are spread farther apart. Unlike a fox or coyote, which trails in a straight line, a dog's trail is staggered and meandering. The tracks of a large dog may

Disappearing eggs could be the doing of the family dog.

be confused with those of a wolf, although the dog's tracks are generally rounder, and the wolf has proportionately longer toenails and larger middle two front toes. The tracks of a large dog are similar to those of cougar, except that the cat has retractable claws, so toenail marks are rarely visible in its tracks.

BEST DETERRENTS. The best protection against dogs is either an electric fence or a sturdy wire mesh fence with horizontal spacing less than 6 inches, vertical spacing less than 4 inches, and a height of 6 feet. An apron or an electrified offset wire toward the bottom will discourage digging. A second offset toward the top will discourage climbing.

CLUES: Domestic Dog

Scat

Feral dog scat is similar to that of a coyote. The scat of a domestic dog fed commercial dog food is uniform in appearance and lacks hair and bone fragments.

Tracks

front

back

Depending on the size of the dog, tracks vary in size from those of a small fox to those of a large wolf. Dog tracks tend to be slightly rounder than tracks of other canids, and have a larger heel pad. The best clue is that a dog's trail meanders, compared to the more direct trails of other canids.

Canines: A Closer Look

Coyotes are often mistaken for foxes, especially when young, although mature coyotes are larger and stouter, with longer legs, longer ears, a longer muzzle, and a shaggy sandy-gray coat. Compared to a red fox, the coyote lacks black boots and usually lacks the white-tipped tail. Compared to a gray fox, the coyote lacks the black tail stripe and cat-like face. Further, the coyote's bushy tail is relatively short and typically dark-tipped, and it normally hangs down, rather than being carried horizontally, like a fox's.

RED FOX

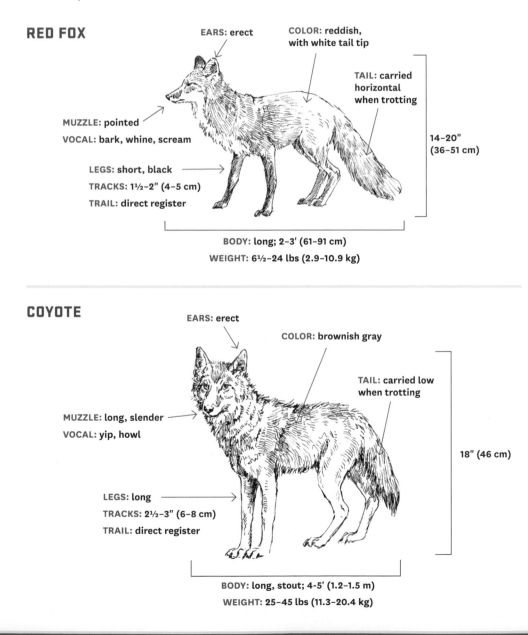

EARS: erect

COLOR: reddish, with white tail tip

TAIL: carried horizontal when trotting

MUZZLE: pointed

VOCAL: bark, whine, scream

LEGS: short, black
TRACKS: 1½–2" (4–5 cm)
TRAIL: direct register

14–20" (36–51 cm)

BODY: long; 2–3' (61–91 cm)
WEIGHT: 6½–24 lbs (2.9–10.9 kg)

COYOTE

EARS: erect

COLOR: brownish gray

TAIL: carried low when trotting

MUZZLE: long, slender
VOCAL: yip, howl

LEGS: long
TRACKS: 2½–3" (6–8 cm)
TRAIL: direct register

18" (46 cm)

BODY: long, stout; 4–5' (1.2–1.5 m)
WEIGHT: 25–45 lbs (11.3–20.4 kg)

THE GRAY WOLF has a broad muzzle, large nose, and relatively small rounded ears, compared to a coyote's narrow, pointed muzzle, small nose pad, and long, pointed ears. The wolf has much larger feet that leave tracks about twice the size of a coyote's.

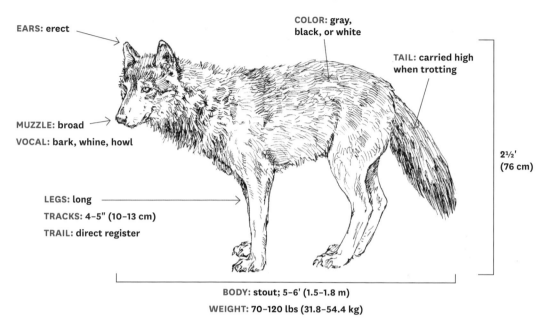

EARS: erect

COLOR: gray, black, or white

TAIL: carried high when trotting

MUZZLE: broad
VOCAL: bark, whine, howl

2½'
(76 cm)

LEGS: long
TRACKS: 4–5" (10–13 cm)
TRAIL: direct register

BODY: stout; 5–6' (1.5–1.8 m)
WEIGHT: 70–120 lbs (31.8–54.4 kg)

FERAL DOGS can be difficult to distinguish from domestic dogs. As a general rule, a feral dog is more aggressive when approached by a human, and it is less likely than a domestic dog to wear a collar.

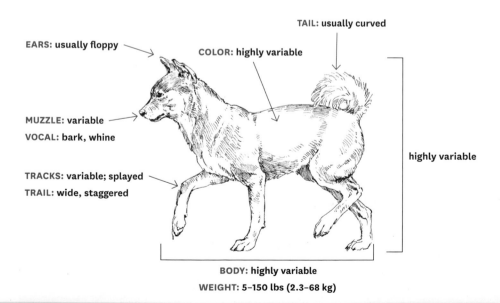

TAIL: usually curved

EARS: usually floppy

COLOR: highly variable

MUZZLE: variable
VOCAL: bark, whine

highly variable

TRACKS: variable; splayed
TRAIL: wide, staggered

BODY: highly variable
WEIGHT: 5–150 lbs (2.3–68 kg)

WEASEL FAMILY

Long-Tailed Weasel · Short-Tailed Weasel · Least Weasel ·
Mink · Marten · Fisher · Otter · Badger · Wolverine

Long-Tailed Weasel

AVERAGE LENGTH (including tail): 22 inches (56 cm)
AVERAGE HEIGHT AT SHOULDER: 3 inches (7.6 cm)
AVERAGE WEIGHT, FEMALE: 6.5 ounces (184 g)
AVERAGE WEIGHT, MALE: 13 ounces (369 g)

Long-tailed weasels (*Mustela frenata*) may be found nearly throughout the contiguous United States, but not much farther north than the Canadian border. The alert, agile, aggressive long-tailed weasel often brings down prey much larger than itself to satisfy its need to maintain a high metabolic rate by consuming the equivalent of up to 40 percent of its body weight each day.

Accordingly, long-tails forage year-round, night and day — more at night during winter and during daylight in summer. Traveling as much as 7 miles in a single hunting excursion, a weasel explores constantly, slithering down tunnels, climbing trees, and swimming streams in its perpetual quest for food.

Each trek, however, entails considerable risk, for weasels in turn are hunted by bobcats, lynx, domestic cats, coyotes, foxes, dogs, raptors, snakes, and other weasels. But long-tails are not the preferred food of any predator, since they are hard to catch, thanks to their speed, nimbleness, and a few evasion tricks.

For one thing, the tip of a long-tail's tail is black, which potentially draws attention away from the animal's head to the less vulnerable appendage. In the North, long-tails develop a white winter coat that blends with the snow — but the tail tip remains black. A second defense is a strong, foul odor released by the weasel's anal gland.

The weasels' dietary preference is for rodents, although long-tails will eat whatever they can find or catch: rabbits, squirrels, gophers, lizards, snakes, frogs, earthworms, insects, fruit, eggs, birds, and even carrion if it comes to that. When they can't find anything else, they'll eat their own siblings or offspring. Their murderous reputation is enhanced by

The long-tailed weasel is long and thin, with short legs and a long tail.

their habit of caching surplus against shortfalls. One small weasel can wipe out an entire poultry flock as quickly as a frenzied red fox.

TYPICAL SIGNS. The weasel's appetite for rodents, and the rodents' appetite for chicken feed, is often what attracts weasels to the coop. A weasel can squeeze its long, thin body through an opening as narrow as 1 inch in diameter, and will hunt mice and rats by traveling through the rodents' own tunnels. In doing so the weasel may eventually find itself inside the coop, especially after having decimated the resident rodent population.

"The killing of domestic poultry may come only after the rat population around the farmyard is diminished," says G. E. Svendsen in *Wild Mammals of North America*. "In fact, rats may have destroyed more poultry than the weasel. In most cases, a farmer lives with weasels on the farm for years without realizing that they are even there, until they kill a chicken."

But once the weasel gets in there, boy howdy — watch out! If you find multiple chickens dead — perhaps piled together or partially hidden — and you feel no predator could possibly have gotten into that coop, it was a weasel. The weasel's needlelike canine teeth are designed to kill a rodent by piercing the throat or skull, and they are just as effective on poultry. Often the only part eaten is the head, which contains the nutritious brain. Sometimes the only damage will be the bluish discoloration of bruising in the skin around the head and under the wings; the theory is that weasels prefer not to break the skin of cached meat, which would hasten deterioration.

In the 35 years we have lived on our Tennessee farm, I have seen weasel sign only

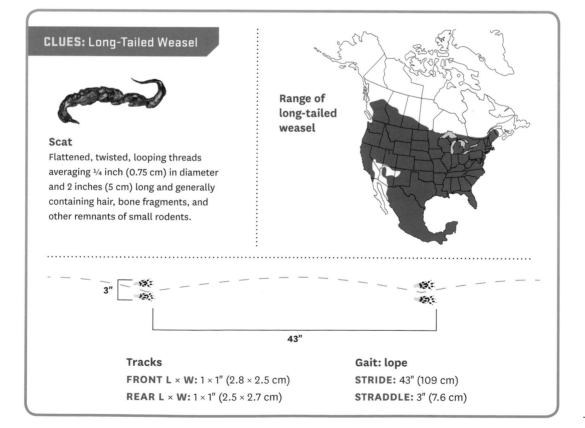

CLUES: Long-Tailed Weasel

Scat
Flattened, twisted, looping threads averaging ¼ inch (0.75 cm) in diameter and 2 inches (5 cm) long and generally containing hair, bone fragments, and other remnants of small rodents.

Range of long-tailed weasel

3"

43"

Tracks
FRONT L × W: 1 × 1" (2.8 × 2.5 cm)
REAR L × W: 1 × 1" (2.5 × 2.7 cm)

Gait: lope
STRIDE: 43" (109 cm)
STRADDLE: 3" (7.6 cm)

once: three piles of fresh scat near the door to our pullet house. The weasel (or weasels) apparently made no attempt to get into the coop, and after I removed the scat I saw no further evidence of a return visit. And that's fine by me, because a weasel confronted by a human can be viciously aggressive.

HABITAT AND RANGE. The long-tailed weasel is found throughout the United States, except in arid areas of Arizona and surrounding states. It readily adapts to a wide variety of habitats including woodlands, thickets, open grasslands, and farmland, wherever small mammal prey and a source of drinking water are available. Each day, a long-tailed weasel drinks nearly 2 tablespoons of water, and therefore it prefers to live near a reliable source of water, such as a stream, river, livestock pond, or drainage canal. Population densities tend to be fairly low and vary with habitat, season, and available resources.

BEST DETERRENTS. Since weasels can weasel through an opening as small as 1 inch, excluding them from a coop presents a definite challenge. For starters, take measures to avoid attracting mice and rats to your coop. To keep out weasels, cover windows, vents, and other openings with ½-inch hardware cloth. If the coop lacks a solid floor to prevent digging, either line the floor with hardware cloth covered with soil or spread an apron fence around all four outside walls. Electric fence, or electric wire offsets around the fence, will prevent climbing.

More Weasels

True weasels belong to the genus *Mustela*, whose name derives from the Latin word *mus*, meaning mouse, in reference to the weasels' dietary preference for rodents. Four species of true weasel inhabit North America: long-tailed weasel, short-tailed weasel, least weasel, and black-footed ferret.

The most likely North American true weasel poultry predators are the long-tailed weasel, the short-tailed weasel, and the least weasel. In some area the ranges of these three overlap. Short-tailed and least weasels are both smaller than long-tailed weasels, less generalized in their dietary preference, and, in the contiguous states, less numerous. Poultry predation signs and deterrents are the same as for long-tailed weasels.

A small and fragmented population of black-footed ferrets (*Mustela nigripes*) is confined to habitat managed by the United States Forest and Wildlife Service, which reintroduced ferrets into western states after the species became nearly extinct. The black-footed ferret dines almost exclusively on prairie dogs, although it will also eat ground squirrels, rabbits, and birds. The species is so rare and so highly managed that the possibility of a black-footed ferret preying on domestic poultry is pretty remote.

A different story, on the other hand, is the pet ferret (*Mustela putorius furo*). Pet ferrets are not related to the native American black-footed ferret but have been introduced from Europe, where they were originally domesticated for hunting rabbits and controlling rodents. If you (or a neighbor) have a pet ferret, and you (or your neighbor) have chickens or young poultry of any kind, a loose ferret may well find its way into the coop and create havoc.

Weasels Are Smaller than You Think

Victoria Redhed Miller, Washington, author, *Pure Poultry*

About a year after we started raising chickens in the foothills of the Olympic Mountains, we experienced our first major predator attack. Our various birds were closed up in their coops at night, so we had assumed they were safe. But eight little chickens, about nine weeks old, were killed inside their coop. Everything looked all right from the outside, but inside the coop was a bloody mess.

The day after the attack, I found out who and how. I was working on a nearby turkey coop when a long-tailed weasel came hopping across the grass, ran up the vertical back wall of the coop where the chickens had been killed, and popped in through a tiny space between wall and roof.

Weasels are much smaller than I thought; picture a jumbo hot dog with a tail, and you'll have a pretty good idea of how small this creature is. I was amazed to learn that a weasel can get through a space as small as 1 square inch.

I began beating up on myself for not learning this fact earlier. Learning from my mistakes is all well — I had built the coop in question — but I felt horrible to learn the lesson at the expense of innocent little lives. I had a hard time not imagining what it must have been like for them — in the dark, not knowing what was happening, only that a destructive force was in the coop, rampaging its way around the roosts until every last bird was dead.

That day, I spent about two hours going over that coop, closing every space that even approached 1 inch. I filled in the space under the hinged corrugated metal roof with horizontal closure strips. I replaced the half-inch hardware cloth floor with solid wood. I screened all ventilation holes with quarter-inch hardware cloth.

Still, the first night we housed a new batch of birds in that coop was nerve-racking. I vowed to learn all I can about building a coop that meets the chickens' needs for safety, as well as health and comfort.

▲ Short-tailed weasel in summer (stoat) and winter (ermine) coat

Short-Tailed Weasel

AVERAGE LENGTH (including tail):
10 inches (25 cm)

AVERAGE HEIGHT AT SHOULDER:
5½ inches (14 cm)

AVERAGE WEIGHT: 2½ ounces (70 g)

Although similar to long-tailed weasels in behavior, habitat, diet, and appearance, short-tailed weasels (*Mustela erminea*) are on average about one-third smaller and have shorter tails proportional to body length. A male, however, can reach twice the size of a female; the male short-tailed weasel is about the same size as a female long-tailed weasel.

Like the long-tailed weasel, the short-tail wears a brown coat in the summer but, unlike long-tails, *all* short-tailed weasels turn white in winter regardless of where they live. The feet are white year-round, and as with long-tails, the tip of the tail is always black. A short-tailed weasel in its summer coat is commonly called a stoat, and in its white winter coat it is known as an ermine.

Short-tailed weasels live in open woodlands, brushy areas, fields, and wetlands, where they hunt mainly at night. They are more proficient swimmers and climbers than long-tailed weasels. Like most predators, they tend to concentrate on whatever prey is abundant and easy to catch.

Least Weasel

AVERAGE LENGTH (including tail): 7 inches (18 cm)

AVERAGE HEIGHT AT SHOULDER: 2½ inches (5 cm)

AVERAGE WEIGHT: 1½ ounces (42 g)

North America's smallest carnivores, least weasels (*Mustela nivalis*) are sometimes called mouse weasels because they are about the size of a mouse — small enough to get caught in a mouse trap, although they rarely are. Since they can squeeze through any opening a mouse can get through (as small as ¼ inch), least weasels are difficult to keep out of poultry housing.

They inhabit much the same range as short-tailed weasels, with the exclusion of the western states. Where their ranges overlap with either short-tailed weasels or long-tailed weasels, the larger species prey on their diminutive cousins.

Least weasels prefer open areas to woodland and are found mostly in fields and marshes. They eat almost exclusively voles and other small rodents but will also prey on small birds and eggs. They are active day and night, year-round, although they are seldom seen by humans. Like the other two species, least weasels leave interesting tracks that wander among nook and cranny and are often irregularly spaced, because weasels typically move by jumping and the distance jumped is not always the same between bounds.

Least weasel

CLUES: Short-Tailed and Least Weasels

Range of short-tailed weasel

Range of least weasel

Scat

The scat of all three North American weasels looks pretty much alike, except the smaller animals leave slightly smaller scat. Color is typically dark brown or black. Look for weasel scat on a log, stump, or rock, or in a latrine near a burrow entrance.

Tracks

The tracks of a short-tailed or least weasel are about half the size of those left by a long-tailed weasel.

| front | front |
| back | back |

Short-tailed weasel   **Least weasel**

Trail

A trail wider than 2 inches (5 cm) was likely made by a long-tail. A trail width between 1¾ and 1⅛ inches (4.5–3 cm) indicates a short-tail. A trail less than 1⅛ inches (3 cm) wide was made by a least weasel.

More Mustelids

The weasel family (Mustelidae) includes not only true weasels but also badgers, fishers, martens, minks, river otters, and wolverines. The family name is a combination of *Mustela* (the genus of true weasels) plus *idai*, a Greek word referring to offspring. How closely related these other species are to true weasels remains to be determined, as they represent quite a degree of diversity.

Mustelids are all aggressive hunters that travel widely in search of food and often attack prey larger than themselves. They don't hibernate and are active year-round, are primarily nocturnal, and are mostly solitary hunters – poultry predation typically involves a single individual. Laws vary from state to state as to which mustelids are protected, which may be seasonally trapped as fur-bearing animals, and which have no legal status. Controlling rodents is the most effective and long-lasting measure for deterring poultry and egg predation by mustelids.

Weasels: A Closer Look

LONG-TAILED

22" long. Brown above, white to deep yellow below. Brown tail with black tip. Feet brownish. Tail 75% of body length.

SHORT-TAILED

10" long. Dark brown above, white below. Brown tail with black tip. Tail 25% of body length.

LEAST

7" long. Dark brown above, white below. Tail short and all brown. Tail 12% of body length.

Mink

AVERAGE LENGTH (including tail):
22 inches (56 cm)
AVERAGE HEIGHT AT SHOULDER:
5 inches (13 cm)
AVERAGE WEIGHT: 2¼ pounds (1 kg)

Although American minks (*Neovison vison*) are more like true weasels than any other mustelid, they differ from weasels in several ways. They have a bushier tail and are slightly longer than the largest true weasel, but much stouter and heavier. Mink fur is a rich dark brown, almost black, with white patches on the throat, chin, and stomach. Unlike some weasels, minks remain the same color year-round.

Minks are most active at night and early morning, although they sometimes hunt during the day. Their diet consists primarily of animals associated with water: muskrats, fish, crayfish, frogs, snakes, and snails. They also eat insects, earthworms, rodents, rabbits, eggs, ducks, and chickens.

Early in my chicken-keeping enterprise, I met a man with an enviable collection of bantams housed in a connected series of covered pens in a wooded area uphill from a creek. One day a mink discovered the pens and, with repeat visits, ended my friend's bantam-keeping venture. The behavior of a mink is similar to that of a weasel, killing multiple birds with bite marks on the head and upper neck, and eating mainly the heads. Waterfowl may be pulled underwater and drowned. Deterrents are the same as for weasels.

CLUES: Mink

Range of mink

Scat
Mink scat looks like weasel scat but is slightly larger, having an average diameter of ⅜ inch and length of 2½ inches (0.9 × 6.4 cm). It is typically found near water, deposited on a log, rock, or clump of grass or weeds.

Tracks
FRONT L × W: 2 × 1½" (5 × 3.8 cm)
REAR L × W: 1¼ × 1¼" (3.2 × 3.2 cm)

Gait: walk
AVERAGE STRIDE: 11" (28 cm)
AVERAGE STRADDLE: 2⅞" (7.3 cm)

front

back

Marten

AVERAGE LENGTH (including tail): 22 inches (56 cm)
AVERAGE HEIGHT AT SHOULDER: 5 inches (13 cm)
AVERAGE WEIGHT: 1¾ pounds (794 g)

About the same size as minks, American martens (*Martes americana*) have catlike ears and a pale chest patch that minks lack. A marten has a grayish head and long, shiny reddish-brown fur that's lighter on the back than on the legs and tail, both of which are nearly black. The bushy tail makes up about one-third the total length.

Martens can rotate their hind feet 180 degrees — allowing them to ascend or descend trees forward or backward — and have sharp, curved semi-retractable claws used for climbing and killing prey. Relatively large footpads allow them to walk atop deep snow. They are

so light-footed that their tracks are rarely seen except in snow, typically leading to or from a tree. A marten will often jump from a tree rather than climb all the way down, leaving a large body print in the snow.

Martens inhabit most of Alaska and Canada, parts of the Rocky Mountains and the West Coast, and isolated areas in the East and upper Midwest. They prefer conifer and mixed hardwood forests, giving them their common name — pine marten. Although they are largely arboreal, they prefer to hunt on the ground.

Martens are mainly crepuscular hunters, although they tend to be more diurnal in the summer months and more nocturnal during the winter, when they are less active. Their winter diet consists chiefly of snowshoe hares, squirrels, or rodents, although they will eat carrion. In the summer, martens enjoy a more diverse diet that includes insects, fruits and berries (unlike the scat of other mustelids, marten scat may contain berry seeds), nuts, birds, and eggs.

Given the opportunity, a marten will eat whole chicks or ducklings, but it is more likely to leave mature poultry decapitated. A marten can easily dig under or climb over a fence that lacks an apron at the bottom and either electrified scare wires or extension arms at the top. A marten can get through an opening as small as 2 inches in diameter, and it can use a tree within 10 feet of a fence to access a poultry yard that lacks overhead protection.

Fisher

AVERAGE LENGTH (including tail): 37 inches (94 cm)
AVERAGE HEIGHT AT SHOULDER: 12 inches (30 cm)
AVERAGE WEIGHT: 7¾ pounds (3.5 kg)

Found only in North America, fishers (*Martes pennanti*) are sometimes called fisher cats, although they aren't cats and they rarely eat fish.

They share much of the same geographic range as martens but don't extend as far north

Range
of marten

into Canada and Alaska, or as far south into New Mexico. Like martens, fishers prefer coniferous and mixed forests, but as their numbers increase they are becoming adaptable to deciduous forests and sometimes even farm areas and suburban backyards, where they can wreak havoc on domestic poultry.

Fishers look like martens but are somewhat bigger — a small fisher female is about the size of a large marten male — and their long fur makes them look bigger yet. A fisher has a longer, pointier muzzle with shorter, wider, rounded ears compared to the marten's more cat-like face and ears. The fisher's coat is dark brown with black legs and tail, while the head is lighter in color. Older males may be grizzled on the back and shoulders. Like martens, fishers have short legs with large, well-furred paws that allow them to walk on snow, sharp semi-rectractable claws for climbing and killing, and reversible hind feet for climbing. Like martens, fishers often leave a tree by jumping down, leaving a body print in snow.

Fishers and martens are the only midsize predators that are both at home in trees and able to squeeze down into tunnels and other small openings seeking prey. Like martens, fishers are solitary, mostly crepuscular hunters, although they may be active during the day in remote areas, as well as in populated areas where they adapt to the presence of humans. Their diet is similar to that of martens, consisting mostly of rodents and rabbits, but also carrion, insects, berries, fruit, birds, and eggs.

A fisher will climb, dig, or chew through wood to gain access to poultry. When it succeeds, it may kill numerous birds, carrying some away (unlike a marten, which is too small to make off with mature poultry). Birds left behind may be missing the head and neck, and sometimes part of the breast. A mama fisher feeding young will carry away a chicken without leaving so much as a feather.

CLUES: Fisher

Tracks
FRONT L × W: 3 × 3⅛" (7.6 × 7.8 cm)
REAR: 2⅝ × 2½" (6.7 × 6.4 cm)

Gait: walk
AVERAGE STRIDE: 18½" (47 cm)
AVERAGE STRADDLE: 5¾" (14.6 cm)

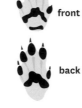 front

back

Range of fisher

Scat

Fisher scat looks like mink scat but is usually a little larger, having an average diameter of ½ inch and a length of 3½ inches (1.2 × 8.9 cm). Unlike mink scat, the ends may be folded and the scat may contain porcupine quills.

WEASEL FAMILY

Otter

AVERAGE LENGTH (including tail):
3½ feet (1 m)

AVERAGE HEIGHT AT SHOULDER:
10 inches (25 cm)

AVERAGE WEIGHT: 16½ pounds (7.5 kg)

River otters (*Lontra canadensis*) are distinct from sea otters (*Enhydra lutris*), fellow mustelids that are more than twice the size, inhabit salt water, rarely come on land, and have no interest in poultry. River otters are more like minks. Both are land mammals that live near aquatic habitats and have webbed feet and a waterproof coat. The otter, however, is much larger than a mink and therefore a more powerful predator. On the other hand, otters are less common and more secretive than minks, shying away from human presence where a mink might boldly go about its business.

A river otter has a long, sleek, streamlined body, short legs, and a long, tapered tail that is never bushy, even when dry. Somewhat seal-like, it has a rounded, slightly flattened face, slightly protruding eyes, and long, stiff whiskers. Its short ears and its nostrils close when it dives into water. The dense, short fur is a rich dark brown on the back and silvery on the belly.

River otters are about the same size as fishers, and their tracks are also similar. Fishers often nose around aquatic habitats, while otters sometimes travel over land, especially young ones seeking to establish their own territories. However, when confronted, the otter will head for water while the fisher will head for the trees. The otter's claws are so short that its track appears to have pointed toes. The otter trail may include tail drag, and on a slope covered with snow, mud, dirt, sand, or grass the trail is likely to turn into a foot-wide body slide.

Otter tracks always lead to or from water. Further, otter scat — unlike dark and well-formed fisher scat, which may include porcupine quills — is green and shapeless, has the texture of oil, smells like fish, and may include bits of shell, fish bones, and fish scales.

River otters eat mainly fish, crayfish, turtles, snakes, and amphibians. They also relish muskrats, eggs, and birds, including ducks and geese — especially any that are young, injured, or molting — and occasionally chickens.

River otter snacking on a baby chick

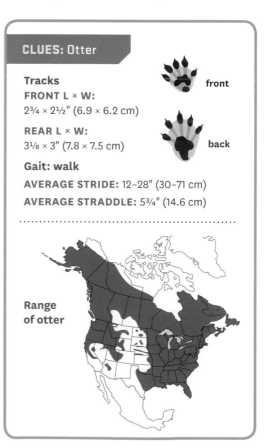

CLUES: Otter

Tracks
FRONT L × W:
2¾ × 2½" (6.9 × 6.2 cm)

REAR L × W:
3⅛ × 3" (7.8 × 7.5 cm)

front

back

Gait: walk
AVERAGE STRIDE: 12–28" (30–71 cm)
AVERAGE STRADDLE: 5¾" (14.6 cm)

Range
of otter

Badger

AVERAGE LENGTH (including tail):
26 inches (66 cm)

AVERAGE HEIGHT AT SHOULDER:
15 inches (38 cm)

AVERAGE WEIGHT: 20 pounds (9 kg)

The badger (*Taxidea taxus*) is a low-slung animal with a small, wide head; long nose; small, round ears; short, thick neck; stout body; short, muscular legs; and short, bushy tail. It has shaggy brownish fur, black legs, and a distinctively striped black-and-white head.

American badgers are known as digging machines. Their front feet have long, backward-curved claws able to excavate like little backhoes, while the smaller rear feet have shorter claws to 'doze away the loosened soil. The badger's lifestyle is based on its ability to rapidly dig up and dine on ground-dwelling rodents. Accordingly, badgers tend to inhabit open areas with friable soil, few trees, and lots of prey — habitats that are abundant in, and mostly west of, the Great Plains from southern Canada to southern Texas

Badgers are solitary and primarily nocturnal and crepuscular, preferring to snooze in an underground den during daylight, especially in hot weather. Since they avoid populated areas and are mainly active after dark, their presence is more often detected by spotting a burrow than a badger.

A badger burrow may be identified by its 10-inch-wide elliptical entrance with a mound of loose soil fanning out from the opening. Look for badger dens along hedgerows, old fences, and the tree line where field meets forest.

Besides eating rodents, badgers relish frogs, snakes, worms, insects, grubs, roots, and anything else they dig up. They also eat turtle and bird eggs, including the eggs of domestic poultry. Among predators that snack on poultry eggs, a badger paws around and leaves the messiest nest. The badger will also prey on poultry and waterfowl, often killing several and burying uneaten parts nearby.

Entry into a poultry yard is gained by digging under a fence or under a wall into a coop's dirt floor. Serious digging is the main sign of badger predation.

STATE ANIMAL

The badger (*Taxidea taxus*) is an official state animal for Wisconsin, long called the "Badger State" after 1800s miners whose tunneling in search of lead ore reminded people of badgers. The state animal was adopted at the urging of elementary school students, who had discovered that the badger had "not been given the official status most people assumed."

WEASEL FAMILY

CLUES: Badger

Range of badger

Scat

Badger scat has a characteristic pungent odor. It may be sausage-shaped but more typically is either folded over or twisted and pointed. Average diameter is ⅝ inch and average length is 4½ inches (1.5 × 11.4 cm).

Tracks

FRONT L × W:
3⅜ × 2⅛" (8.6 × 5.3)

REAR L × W:
2⅜ × 1¾" (5.9 × 4.3)

front

back

Gait: walk

AVERAGE STRIDE:
15" (38 cm)

AVERAGE STRADDLE:
6¼" (16 cm)

Badger scat is not a reliable sign because it is so variable in shape and size that it is easy to mistake for that of other carnivores, and it is often buried and therefore not visible. Badger tracks can be mistaken for those of a skunk or a coyote. Like a badger, a skunk has long claws on the front feet, although the tracks are considerably smaller than a badger's. Compared to a coyote's track, the badger's track is pigeon-toed and has longer claw marks. Unlike a skunk or coyote, the badger tends to drag its feet.

Aside from strict rodent control, deterring badgers involves discouraging digging. A concrete coop floor, or a wooden floor raised off the ground, prevents digging into the coop. Lining the pen bottom and sides (or creating a 2-foot apron fence) with 10-inch-mesh 14-gauge wire, buried 18 inches deep, should exclude most badgers. A highly effective but short-term measure is to light the yard with high-intensity floodlights.

Wolverine

AVERAGE LENGTH (including tail):
3 feet (91 cm)

AVERAGE HEIGHT AT SHOULDER:
12 inches (30 cm)

AVERAGE WEIGHT: 30 pounds (13.5 kg)

Although wolverines (*Gulo gulo*) are reputed to be killing machines, in fact much of their diet consists of animals that are already dead, having been killed by some other predator, buried in an avalanche, or trapped as a fur bearer. Wolverines scavenge such carcasses because they are not proficient runners. In deep snow, however, a wolverine's snowshoe-like feet give it an advantage over larger animals, such as caribou or moose, that get bogged down and can't escape. In summer months, wolverines fill their bellies with rodents and other small animals, as well as berries, birds, and eggs.

The wolverine is a stocky, muscular animal with long, dense brown fur and a bushy tail. It has short legs and relatively large webbed and furry feet with sharp claws useful for self-defense, climbing, digging, and killing prey. It has a broad, round head with a short muzzle and a powerful jaw able to bite into frozen meat and crack bones. Each wolverine has a distinct pattern of yellowish stripes on its face, neck, and chest by which it may be individually identified.

Most of the wolverines' domain is inhospitable to poultry, and they tend to avoid human settlements. However, their typically food-scarce environment leads them to travel great distances, sometimes into plains or farmland, in search of a meal — which may well be someone's free-range poultry or nesting waterfowl.

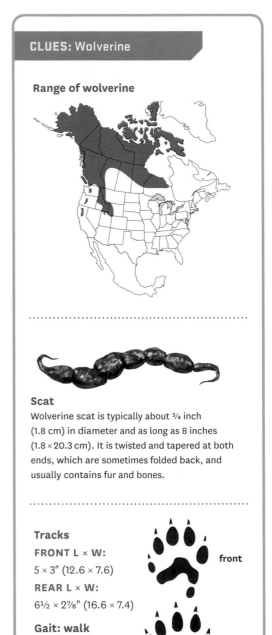

CLUES: Wolverine

Range of wolverine

Scat
Wolverine scat is typically about ¾ inch (1.8 cm) in diameter and as long as 8 inches (1.8 × 20.3 cm). It is twisted and tapered at both ends, which are sometimes folded back, and usually contains fur and bones.

Tracks
FRONT L × W:
5 × 3" (12.6 × 7.6)
REAR L × W:
6½ × 2⅞" (16.6 × 7.4)

Gait: walk
STRIDE:
23" (58.4 cm)
STRADDLE:
8½" (21.6 cm)

front

back

Bobcat

AVERAGE LENGTH: Male, 3 feet (.91 m); female, 2 feet (.61 m)

AVERAGE HEIGHT AT SHOULDER: 15 inches (38 cm)

AVERAGE WEIGHT: Male, 25 pounds (11.3 kg); female, 17.5 pounds (7.9 kg)

Bobcats (*Lynx rufus*) are found throughout most of the United States and in the southern part of Canada. Their preferred diet is rabbits, but they will eat whatever they can easily catch, which includes chickens, ducks, turkeys, and eggs. They usually hunt at dawn and dusk but may be active during the day in rural and isolated areas. In urban areas, they avoid contact with people by prowling mostly at night.

Bobcats are quiet and reclusive and therefore are rarely seen. On my farm, we occasionally see bobcat tracks in mud or snow, but only once in several decades did we see an actual cat. It wasn't until we set up a trail cam that we realized how much time bobcats spend roaming our farm. Each cat frequents a predictable route, usually following an existing trail or roadway.

Most of the time a bobcat travels alone, or a female may be accompanied by her kittens. The cat hunts by sight and sound, spending a lot of time just sitting or crouching in one place, swiveling its head and ears to look around and listen. Once a bobcat spots prey, it will creep within range and then make a fast dash and pounce. A bobcat can handily climb up a tree or fence post to get inside a poultry yard and can jump over a fence that's 6 feet high or even taller.

TYPICAL SIGNS. A typical sign of a bobcat attack is a single bird missing. You might see a patch of feathers where the bird struggled as it was caught, or a trail of feathers where the body was dragged away. Typically a bobcat will find a place to hide its catch, cover it with leaves, grass, or snow, and frequently come back for a snack until most of bird has been eaten. The cat is then likely to look for another one. If you find the bird's body, most likely the head will have been eaten and you'll see claw marks on the neck, back, and sides.

Another way to confirm the presence of a bobcat is to identify droppings, tracks, or claw marks (called rakes or scrapes) on trees. A bobcat usually covers its droppings with soil, leaves, grass, snow, gravel, or whatever is handy. Scratch marks around covered scat extend 12 to 18 inches in all directions.

Like a house cat, a bobcat marks its territory by leaving scrapes on fence posts, stumps, and trees. Compared to the 1½- to 2-foot-high scratchings of a house cat, the bobcat scratches 2 to 3 feet above ground.

APPEARANCE. The bobcat gets its name from its short bobbed tail, which is only 4 to 6 inches long. The average bobcat is only about twice the size of a typical house cat. The male weighs 20 to 30 pounds and its body is about 3 feet long. The female is somewhat smaller, weighing 15 to 20 pounds, and has a body length of about 2 feet. Height at the shoulder ranges from 12 to 18 inches. In cold climates bobcats tend to be larger than those in warm areas.

HABITAT AND RANGE. Bobcats like cliffs and ledges that offer protection and provide shelter for their kittens. In a wooded area, a bobcat may take up residence in a large pile of brush or logs, or in a hollow tree or log. It will travel through an open field if it can quickly retreat to nearby dense brush or forest. Bobcats can be attracted to urban areas, where their young are safer from coyotes and the hunting is good thanks to an abundance of citified rodents.

Depending on available resources for hunting, secure places to hide, and the local population density of bobcats, a female's territory may extend for up to 3 square miles. A male's territory is about twice that and overlaps the territories of two or three females. Territory boundaries are marked by scrapes, scent, and scat. A bobcat that is trapped and moved will do its best to return to its home territory.

BEST DETERRENTS. Being basically reclusive, a bobcat usually can be easily persuaded to steer clear. A noisy radio and/or motion-activated lights may make it uncomfortable enough to move on, at least until the cat realizes they are not a threat. Since bobcats look for easy prey, the best way to protect poultry from them is to secure the birds at night.

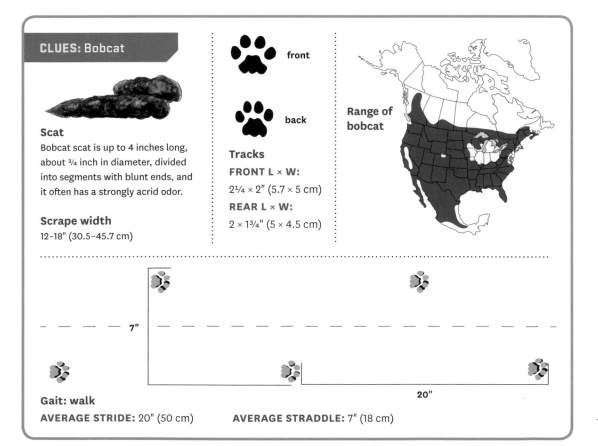

CLUES: Bobcat

Scat
Bobcat scat is up to 4 inches long, about ¾ inch in diameter, divided into segments with blunt ends, and it often has a strongly acrid odor.

Scrape width
12–18" (30.5–45.7 cm)

front

back

Tracks
FRONT L × W:
2¼ × 2" (5.7 × 5 cm)
REAR L × W:
2 × 1¾" (5 × 4.5 cm)

Range of bobcat

7"

20"

Gait: walk
AVERAGE STRIDE: 20" (50 cm)   **AVERAGE STRADDLE:** 7" (18 cm)

The Case of the Vanishing Turkeys

The first Royal Palm turkey that disappeared from our farm vanished without a trace. Several days later, another turkey disappeared.

We had hired a contractor to do some work behind the barn. A job that should have taken a few days dragged on for weeks thanks to February rain followed by freezing weather and snow. The barnyard was in chaos. Disturbed soil left gaps under gates and along fence lines, and each evening we had to patrol for electric fences turned off and gates left open.

At least the snow gave me a chance to look for predator tracks. I shot photos of various crisscrossing trails. Most of the tracks were made by critters too small to carry a turkey. However, one track squished in the snow looked like it might have been made by a dog. Or a big cat. I compared it to photos of our cat's print in the snow. Apart from size, the two prints were nearly identical.

We set up a trail cam in the turkey yard. The snow melted and another turkey disappeared. This time, we found a trail of white feathers leading into the woods. We located the recently missing turkey buried in leaves under a rock outcrop. Sign of a bobcat.

Checking the trail cam, we finally saw our perp. It was indeed a bobcat. But how was the cat getting inside the electric fence? We repeatedly walked the fence line without spotting any problems, and our fence tester indicated that the wires were plenty hot.

We bought a fence monitor and as soon as we clipped it to the fence surrounding the turkey yard it started flashing. Again we walked up and down the fence line. Eventually we saw that a ground wire had been loosely wrapped around a hot wire near the back gate, where the contractors had been going in and out. Oddly, it hadn't completely grounded the fence, nor did it arc. With no audible snapping sound or visible spark, the problem had been difficult to find. But find it, and fix it, we did.

A few days later we were inside the barn at nightfall when we heard the startled yowl of a big cat coming from the darkest back corner of the fence. Our guess is the cat had planned to again slip into the yard when it got a big fat zap on the nose. Since then our trail cam has captured a bobcat occasionally passing by the barn, but so far it has shown no inclination to venture inside the fence.

Comparing Cat Shape and Size

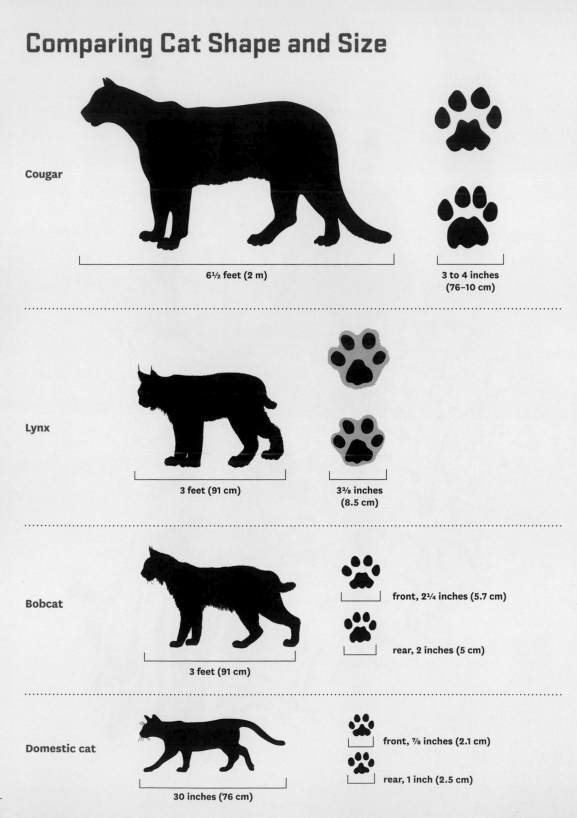

Cougar

6½ feet (2 m)

3 to 4 inches
(76–10 cm)

Lynx

3 feet (91 cm)

3⅜ inches
(8.5 cm)

Bobcat

front, 2¼ inches (5.7 cm)

rear, 2 inches (5 cm)

3 feet (91 cm)

Domestic cat

front, ⅞ inches (2.1 cm)

rear, 1 inch (2.5 cm)

30 inches (76 cm)

CAT FAMILY

More Cats

The cat family (Felidae) consists of two subfamilies: big roaring cats (Pantherinae), such as lions, tigers, and jaguars, and smaller purring cats (Felinae). Except for an occasional vagrant jaguar along the Mexican border, the four cat species found in the United States and Canada are all Felinae — bobcats, cougars, lynx, and house cats. Some areas of southern Canada and northern United States are home to all three wild species.

Members of the subfamily Felinae, commonly known as felines, cannot roar like the big cats. However, like big cats, felines make a living killing and eating other animals, an occupation for which their teeth are ideally suited. Their long canine teeth are designed for grasping and killing prey, and their sharp molars can cut through meat and bone.

A feline's back legs are longer and more powerful than the front legs, giving it strength for leaping and pouncing, and for short bursts of speed. Sharp claws are used for gripping prey, climbing, and fighting. The claws are retractable, which keeps the nails sharp, unlike the nonretractable nails of a dog or bear that get worn down and blunted. Each paw has an M-shaped pad and four toes, with nails typically not visible in the tracks. As toe walkers, cats can move fast.

All cats have a natural righting reflex, which is the ability to rotate their bodies so they land on all four feet when they fall. The cat turns its head first, and then its flexible spine, thanks to 30 vertebrae (compared to a human's 24). The cat then extends its legs downward so that its leg joints will absorb the impact of landing. The righting reflex is handy for an animal that frequents trees and other high places.

Felines are solitary hunters, with the exception of a mother training kittens. Toms, which are larger than females, join females only during breeding season. The mother alone cares for the offspring.

Like all predators, cats play an important ecological role by keeping their prey species in check. Our three North American felines, in addition to bobcats, are described here from largest to smallest.

STATE ANIMAL

The Florida panther (*Felis concolor coryi*) is the official state animal in Florida, the only state inhabited by this rare and endangered cougar subspecies.

The bobcat (*Lynx rufus*) is New Hampshire's official state wildcat, selected at the urging of fourth- and fifth-grade students, who described the animal to state legislators as "strong, resilient, independent and adaptable. Bobcats are just like the people of this Granite State."

Cougar

AVERAGE LENGTH (including tail): 6½ feet (2 m)

AVERAGE HEIGHT AT SHOULDER:
25 inches (64 cm)

AVERAGE WEIGHT: 130 pounds (59 kg)

A slender, muscular, long-bodied cat, the cougar (*Puma concolor*) is the largest feline. Some individuals can weigh more than the smaller of the so-called big cats. A mature male can weigh as much as 200 pounds. On average, females weigh about one-third less than males. At maturity, the cougar's tail is 20 to 30 inches long, nearly the same length as its head and body combined. The cougar goes by many names, including catamount, mountain lion, and puma.

Cougars inhabit primarily wilderness areas in the western mountains of Canada and the United States, although their range has been steadily expanding eastward as individuals seek new territory. They are not fussy about habitat, as long as it offers dense cover for mounting sneak attacks, along with deer or other large species to ambush. Each cougar eats the equivalent of one deer every two weeks, an appetite requiring a large home range.

When large prey like deer or elk are scarce, a cougar will eat mice, squirrels, rabbits, beavers, porcupines, raccoons, skunks, martens, coyotes, bear cubs, bobcats, lynx, and large birds. A cougar is unlikely to visit your poultry yard unless the snowpack is deep, hunting is poor, or kittens are hungry. A cougar won't tear through a fence or dig underneath but will climb or leap to get into a poultry yard. A cougar can climb a 12-foot fence or leap 15 feet straight up into a tree or onto a roof.

CLUES: Cougar

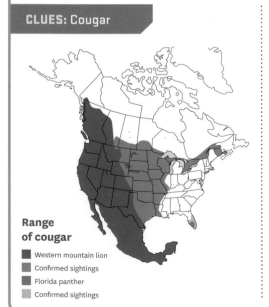

Range of cougar

- Western mountain lion
- Confirmed sightings
- Florida panther
- Confirmed sightings

Scat

Cougar scat is segmented and 1 inch (2.5 cm) or more in diameter and as much as 17 inches (43 cm) long, and it usually contains hair and bone bits that give it the appearance of white charcoal briquettes. Look for scat under a big tree, at trail intersections, in the middle of a trail, along a ridge, or near cached prey.

Scrape width
20–36" (51–91.5 cm)

Tracks

A cougar's tracks are round, 3 to 4 inches (7.6–10.2 cm) long, and 3 to 4½ inches (7.6–11.4 cm) wide. The size and shape are similar to a wolf or large domestic dog track.

front

back

Nothing short of a sturdily constructed coop that can be closed up at night, with a securely covered run surrounded by an electric fence or a chain-link fence with extension arms, will protect poultry from a determined cougar. Bright security lights serve as an additional deterrent, as do motion- or timer-activated noise, lights, or water spray. Do not feed deer or other wildlife, and clean up windfall orchard fruit that attracts deer, which in turn attract cougars.

Lynx

AVERAGE LENGTH (including tail): 3 feet (91 cm)
AVERAGE HEIGHT AT SHOULDER: 20 inches (52 cm)
AVERAGE WEIGHT: 26 pounds (12 kg)

Closely related to a bobcat, and sometimes confused with it, a lynx (*Lynx canadensis*) is slightly bigger than its cousin but appears to be much larger because of its longer hair. Other key differences are that the lynx is mottled more than spotted and has longer tufted ears, a shorter stubby tail, longer legs, and larger furry paws. Both species are called wildcats. The lynx is also called the Canada or Canadian lynx to distinguish it from the bobcat, which is sometimes called the common lynx.

With a less varied diet than the bobcat's, the lynx has a predilection for snowshoe hares (*Lepus americanus*). Snowshoe hares inhabit deep snow country southward from the arctic tree line, through much of Alaska and Canada, into parts of the northern states and down the Rocky Mountain range. The best way to know if lynx are likely to prowl in your area is to

determine if snowshoe hares live there. Each lynx eats approximately 200 hares per year; the more hares, the more lynx.

Snowshoe hare populations follow a roughly 10-year boom-and-bust cycle. When hares are scarce, and in the less-snowy southern areas that support fewer hares, the lynx's diet branches out to encompass rodents and other small mammals, fish, birds, and eggs.

Poultry predation may occur in autumn, as well, when a mama lynx takes her kittens on family hunting forays. Poultry are typically bitten in the head and neck, and the heads are often eaten. Deterrents are the same as for bobcats and cougars.

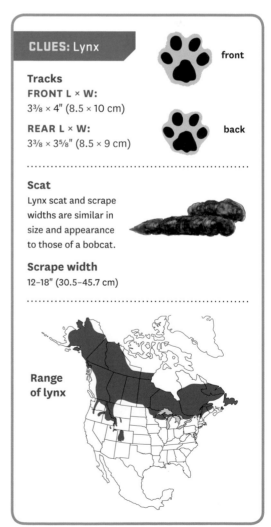

CLUES: Lynx

Tracks
FRONT L × W:
3⅜ × 4" (8.5 × 10 cm)

REAR L × W:
3⅜ × 3⅝" (8.5 × 9 cm)

front

back

Scat
Lynx scat and scrape widths are similar in size and appearance to those of a bobcat.

Scrape width
12–18" (30.5–45.7 cm)

Range of lynx

House Cat

AVERAGE LENGTH (including tail):
30 inches (76 cm)
AVERAGE HEIGHT AT SHOULDER: 10 inches (25 cm)
AVERAGE WEIGHT: 10 pounds (4.5 kg)

Watching kittens play gives you a pretty good idea of how the wild Felinae kittens learn their trade. They spar, ambush, and chase each other and climb on curtains and furniture. In doing so, house kittens learn the same skills that wild cats develop to successfully hunt for prey and elude predators. House cats allowed outdoors instinctively hunt small animals — rodents, reptiles, amphibians, and birds — even when they're not hungry.

A lost or abandoned house cat is able to survive because it's nearly as good at hunting as its larger wild cousins. Kittens born to a stray house cat become feral cats, commonly called alley cats. Because they have never been handled, they typically avoid contact with humans and live just as independently as any other wild feline.

Feral cats readily adapt to just about any climate and are becoming more plentiful in both the United States and Canada. In a safe habitat where hunting is good, a feral cat can give birth to 2 to 10 kittens up to three times a year, beginning at about six months of age. In urban and suburban areas, where rodents and garbage are plentiful, groups of cats may live in colonies of up to 15 individuals. In rural areas

they tend to be more solitary. Most feral cats are somewhat smaller than the average domestic cat. They usually have short hair and can be any of the same colors as domestic cats.

Like other small predators, feral cats enjoy eating eggs and poultry up to about the size of an average chicken or a duck. Cats are messy eaters and will scatter remains over a relatively large area, where you might find a bird's wings, feet, bones with tooth marks, and skin with feathers attached. Young or small poultry may be eaten whole or carried away. Until a feral cat is barred entry, it will keep coming back for more. House cats, too, may take an interest in poultry, but they primarily target chicks, ducklings, and other young ones.

The best way to prevent predation by feral cats is to eliminate cozy places for them to curl up during day. Block off entrances to foundations, close up unused buildings, and remove abandoned machinery and debris piles. Reducing these attractants also helps control rodents, which otherwise would attract feral cats. And don't encourage neighborhood strays and feral cats by feeding them.

CLUES: House Cat

Scat

Feral and domestic cat scat is dark brown, about 3 inches (7.5 cm) long, segmented, and usually buried. Although house cat scat may have some hair (from grooming), feral cat scat typically contains more hair and some small bones.

Scrape width
7–10" (17.7–25.4 cm)

Tracks

FRONT: ⅞ × 1¼"
(2.1 × 3.3 cm)   front

REAR: 1 × 1⅜"
(2.5 × 3.4 cm)   back

A domestic cat will rid the barn of mice but may also prey on chicks and other small poultry.

Cats: A Closer Look

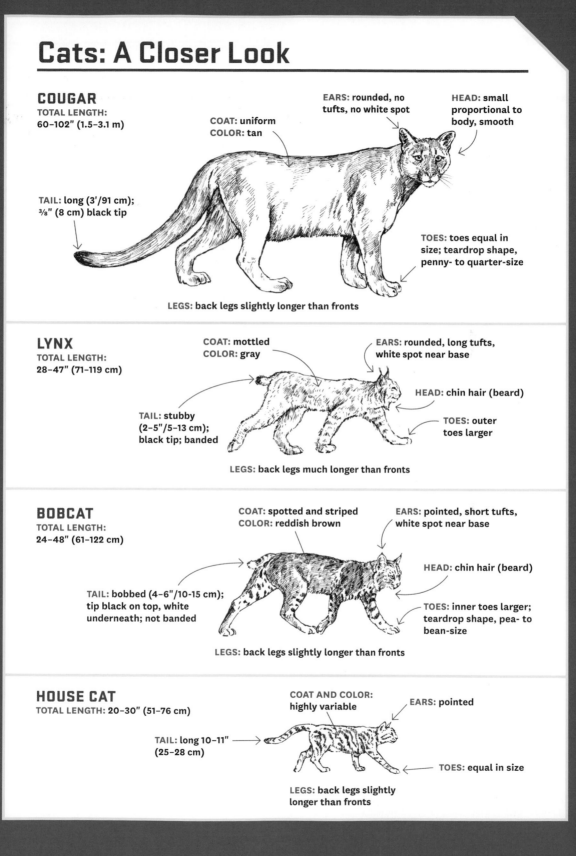

COUGAR

TOTAL LENGTH:
60–102" (1.5–3.1 m)

COAT: uniform COLOR: tan

EARS: rounded, no tufts, no white spot

HEAD: small proportional to body, smooth

TAIL: long (3'/91 cm); 3/8" (8 cm) black tip

TOES: toes equal in size; teardrop shape, penny- to quarter-size

LEGS: back legs slightly longer than fronts

LYNX

TOTAL LENGTH:
28–47" (71–119 cm)

COAT: mottled COLOR: gray

EARS: rounded, long tufts, white spot near base

HEAD: chin hair (beard)

TAIL: stubby (2–5"/5–13 cm); black tip; banded

TOES: outer toes larger

LEGS: back legs much longer than fronts

BOBCAT

TOTAL LENGTH:
24–48" (61–122 cm)

COAT: spotted and striped COLOR: reddish brown

EARS: pointed, short tufts, white spot near base

HEAD: chin hair (beard)

TAIL: bobbed (4–6"/10-15 cm); tip black on top, white underneath; not banded

TOES: inner toes larger; teardrop shape, pea- to bean-size

LEGS: back legs slightly longer than fronts

HOUSE CAT

TOTAL LENGTH: 20–30" (51–76 cm)

COAT AND COLOR: highly variable

EARS: pointed

TAIL: long 10–11" (25–28 cm)

TOES: equal in size

LEGS: back legs slightly longer than fronts

BEAR FAMILY

Black Bear · Brown and Grizzly Bears

Black Bear

AVERAGE LENGTH (including tail): 5½ feet (1.7 m)
AVERAGE HEIGHT AT SHOULDER: 3 feet (.91 m)
AVERAGE WEIGHT: 170 pounds (77 kg)

The commonest, smallest, shyest bear species of North America, black bears (*Ursus americanus*) prefer to avoid humans altogether. Nevertheless, they go where food is plentiful, even if that means wandering into town to dine on garbage. In their natural habitat, black bears typically forage during daylight hours, most actively at dawn and dusk. To avoid humans or brown bears, they become more nocturnal.

Black bears are mainly solitary. Exceptions are during midsummer breeding, when a mother is raising her latest cubs, and when bears congregate at feeding sites where food is abundant, such as an orchard, berry patch, or landfill.

Although bears are classified as carnivores, they are actually opportunistic eaters that, surprisingly, enjoy a primarily vegetarian diet. Maintaining their massive bodies requires lots of calories. They use their keen sense of smell to seek food that's easy to get in quantity.

Some of their favorites are fruits and berries, nuts, plant sprouts, leaves, tree buds, and twigs. Meat makes up only about 15 percent of their diet. Depending on the availability of plant-based foods, a bear may round out its diet by eating carrion, garbage, insects, fish, small mammals, birds, and eggs.

Just because you see a bear on your place doesn't mean it's up to mischief. A bear may travel many miles attempting to fill its stomach and could be just passing through on its way to happier hunting grounds.

TYPICAL SIGNS. The first signs you are likely to see after a bear has visited your poultry yard are a torn-down fence and a knocked-down or ripped-open coop door or window. Birds may be missing, having been moved to a secluded spot for leisurely dining. Remaining birds may be mauled and mutilated, with intestines scattered. You may find bear scat nearby. A bear will generally take several birds at a time. Where poultry continues to furnish easy meals, the bear will almost certainly be back for more.

Bears are most likely to visit areas of human development when natural foods are scarce. The time to watch out for bears is during a lean year when mast (acorns and other forest nuts) is sparse. At such times, bears are attracted by the delectable odors of garbage, compost, corn, pet food, bird seed, and chicken feed. Chowing down on chickens may be just an afterthought.

If you do see a bear, remain at a safe distance, preferably indoors, and make lots of loud noises — slam a door, clang pots, turn up the volume on a radio or TV. Don't go out until you are certain the bear has left. Bear injuries to humans most often occur when a person gets too close. A bear that feels threatened will respond fast and furiously. Double that if the bear is a mother with cubs.

SEASONALITY. Bears eat voraciously from midsummer to fall, fattening themselves in preparation for winter hibernation. During autumn, a black bear forages up to 20 hours a day, packing in as many as 20,000 calories a day and getting them wherever it can find them, including poultry yards.

Where food is scarce — for example, in an area that gets early deep snow — black bears may go into hibernation as early as September.

When food remains readily available, black bears usually begin hibernating in November or December.

In a cold northern climate, black bears generally sleep through the winter. In a more southerly area, where some winter days may be relatively warm, a black bear may wake up and wander around for a bit, and maybe grab a snack, before returning to its den. Bears are more active where winter food is accessible.

Winter hibernation lasts between five and seven months, less in warmer climates and more in colder climates. During hibernation, a black bear's body uses as many as 4,000 calories a day, derived from stored body fat. A black

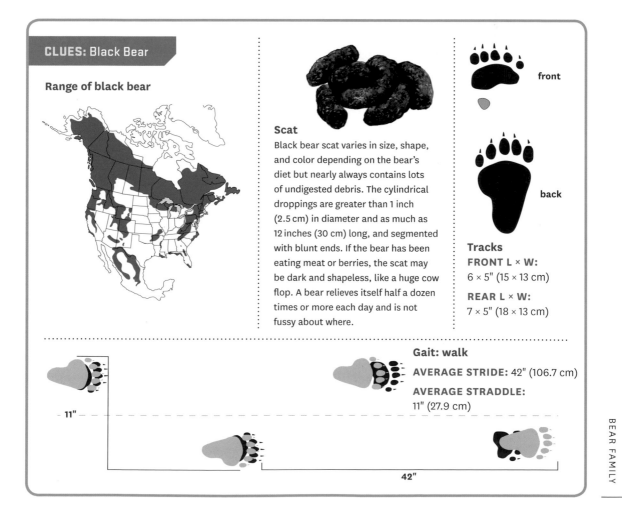

CLUES: Black Bear

Range of black bear

Scat

Black bear scat varies in size, shape, and color depending on the bear's diet but nearly always contains lots of undigested debris. The cylindrical droppings are greater than 1 inch (2.5 cm) in diameter and as much as 12 inches (30 cm) long, and segmented with blunt ends. If the bear has been eating meat or berries, the scat may be dark and shapeless, like a huge cow flop. A bear relieves itself half a dozen times or more each day and is not fussy about where.

front

back

Tracks

FRONT L × W:
6 × 5" (15 × 13 cm)

REAR L × W:
7 × 5" (18 × 13 cm)

Gait: walk

AVERAGE STRIDE: 42" (106.7 cm)

AVERAGE STRADDLE:
11" (27.9 cm)

11"

42"

Bear Attack!

Pam Art, Massachusetts, former president of Storey Publishing

Early one August evening, we returned to our house after spending a wonderful few hours at my retirement party. While I relaxed in the living room chatting with a friend, my husband Henry went to close the chickens in their coop. We had five, about three months old, living in a prefab Eglu coop-with-run. This was our second batch of chicks growing up in the Eglu and we had been pleased with its functionality and design.

As soon as Henry was out the back door in the growing dusk, he saw the coop was turned upside down and a big black bear was going after the feed and stomping on the trapped chickens. I heard his tremendous shout "NO!" and got to the door in time to see Henry chase the bear into the nearby field.

Henry came back holding a dead chicken the bear had dropped. Sadly, all but one chick were dead. When we finally got Cleo, a beautiful golden laced Wyandotte, out from her hiding spot under the porch she was uninjured, but it was a couple of hours before she no longer appeared to be in shock.

We have seen bears before but never had any trouble. We believe two things led to this attack. First, it was the season when bears bulk up before hibernation, and eating is their constant preoccupation. Second, we had just switched to a poultry feed with molasses in it and the bear likely was drawn by the odor. After the bear turned over the coop to get at the feed, the chickens became the second course.

Cleo was not going to be a calm, happy chicken all by herself, so Storey's Creative Director, Alethea Morrison, and her family adopted Cleo into their flock. They renamed her Lucky. She is still going strong and, several years later, is still laying. We have a new flock now and no longer buy feed with molasses in it. Bears have visited our new wooden shed coop but haven't been able to budge it, so they continue on their way without bothering our chickens.

bear usually emerges from hibernation around late February, weighing 20 to 30 percent less than in the fall. Yet for the first couple of weeks it eats only lightly while its body adjusts to a regular summer diet.

For the bear, that's a good thing, because natural food sources may not yet be available in early spring. Backyard poultry, on the other hand, are nearly always available, making them particularly attractive at this time. During the summer, a black bear may consume up to 8,000 calories a day to restore its body systems before it once again begins fasting. Luckily for poultry, the summer months offer a resourceful bear plenty of other things to eat.

Toward the end of hibernation, a female black bear gives birth to two or three cubs, on average, every two years or even less often. A group of bears foraging together in the fall will likely be a mother and her most recent cubs. The cubs hibernate with their mother during their first winter and disperse the following spring. A male yearling might have to travel quite a distance to find an unoccupied home range.

APPEARANCE. You would think that a black bear would be black, and indeed most of them are. But in western states, a black bear's coat may be any shade of brown from dark to blond, and in British Columbia a black bear may be bluish gray or even white.

Average sizes range from about 90 to 250 pounds. Males are larger than females and can reach record weights exceeding 800 pounds. The length of a mature bear ranges from 4 to 7 feet from its nose to the tip of its short tail. Height at the shoulders is 2½ to 3½ feet.

A black bear has large feet, powerful legs, and sharp claws. It can climb a tree lickety-split. It can stand erect and even walk a short distance on two legs when it wants to get a better view or a closer whiff, reach for something tasty, or play or fight with other bears. Normally the bear shuffle-walks on all fours, with its big belly hanging nearly to the ground. When motivated, a black bear can sprint as fast as 30 miles an hour.

HABITAT AND RANGE. Black bears inhabit large forested tracts from northern Alaska, throughout most of Canada, into the northwestern United States and along the East Coast down to Florida — the only bears living east of the Mississippi River. A few black bears make their home in limited areas of the Southeast and Southwest.

The home range of a female black bear is between 2½ and 10 square miles, depending on the availability of resources. Except for a mother with young cubs, females do not share territories. The home range of each male is between 10 and 60 square miles. A male's

STATE ANIMAL

The black bear (*Ursus americanus*) is an official state mammal for Alabama, New Mexico, and West Virginia. In Louisiana, the subspecies Louisiana black bear (*Ursus americanus luteolus*) is a state mammal.

The grizzly bear, a common name for the inland-dwelling brown bear (*Ursus arctos*), is an official Montana state mammal. An extinct subspecies, the California grizzly bear (*Ursus arctos californicus*), is a California state mammal, even though the last known grizzly in that state was shot in 1922.

territory may overlap the territories of several females, and the territories of several males may overlap the range of a single female.

Black bears take up residence where they can find ready food resources and a reliable source of drinking water. A bear seen on open farmland or in an urban area that lacks suitable forest habitat is commonly just passing through. As with coat color, size, and weight, travel patterns and specific habitat preferences vary with geographic location, as do hibernation times, tolerance of humans, and other behavior patterns.

BEST DETERRENTS. The best way to discourage black bears is to eliminate such food attractants as a greasy backyard grill, a bird feeder left out overnight, or garbage cans that aren't bear proof. Store poultry feed in a secure building or in a bear-proof container, available in bear country from many hardware stores. The Interagency Grizzly Bear Committee maintains a list of "Certified Bear-Resistant Products" available online (see Resources). Poultry themselves can attract a hungry bear with their odor, which tempts a bear from a long way off.

A common theme in backyard poultry keepers' descriptions of bear attacks is a lack of both adequate fencing and coop security. A quality-built electric fence goes a long way toward keeping bears at bay. To protect chickens and other poultry where bears are a known problem, some local governments offer funding to help pay for electric fencing, and some ordinances provide a variance for erecting an electric fence within city limits.

To keep out bears, the minimum number of electrified scare wires added to an existing fence is three. The minimum number of electrified wires on a perimeter fence is five, with a minimum total height of 40 inches (100 cm). The energizer should have a minimum output of 1 joule and 6,000 volts, and the fence must be

▲ Well-placed electrified scare wires will discourage bears from climbing a fence.

well grounded. A coop completely surrounded by a well-built electric fence is pretty well protected from bears.

Should a bear breach the fence, however, most chicken coops provide woefully inadequate protection. Poultry that live entirely indoors, or at least are secured indoors at night, need a super-strong coop. Environment Canada recommends ⅝-inch plywood and 2 × 4 construction with screws (not nails). Use heavy-duty hinges and slide-bolt latches, mounting hinges inside the building so a bear can't tear them off. Cover siding seams with metal flashing so claws can't pull the siding apart. Mount hinges, seams, flashing, and access doors flush, leaving no corners for a bear to get a claw underneath. Cover windows and vent openings with sturdy hardware cloth, mounted with washers and screws, or better yet, install security welded-wire mesh window guards (a keyword search will yield lots of sources). Locate your coop away from bear attractants such as fruit trees, berry bushes, and compost and in an open area free of thick brush and other potential bear cover.

BEAR FAMILY

The bear family (Ursidae) in North America includes three species: black bear, brown bear, and polar bear. While black bears and brown bears occasionally dine on poultry, polar bears live on arctic sea ice and have a dietary preference for seals, so they are not likely to be much of a threat to domestic poultry. Bears in North America share these features in common:

- They are large, heavy bodied, and clumsy looking.

- Males are larger than females and can weigh twice as much.

- They have heavy paws, each with five toes.

- They have large, sharp, curved, nonretractable claws.

- The rear footprints look like a human's, but with the big toe on the outside.

- Like humans, bears walk flat-footed (plantigrade).

- When motivated, bears can run faster than 30 miles an hour.

- They have heavy heads, strong jaws, and large noses.

- They have small eyes and ears.

- Their sense of smell is better developed than their sight and hearing.

- They can stand erect and walk a few steps on their hind legs.

- They have stubby tails.

- They have long, dense hair.

- They are solitary hunters and foragers.

- They rarely vocalize, other than occasional growls.

- Mature bears have few natural enemies.

BROWN BEAR OR BLACK BEAR?

Despite their many similarities, brown bears and black bears differ in many ways, including temperament. Brown bears are much more aggressive toward humans than black bears are, and they are more difficult to deter.

Comparing Bear Shape and Size

Black bear

4–7' (1.2–2.1 m)

Brown/grizzly bear

5–9' (1.5–2.8 m)

BEAR FAMILY

Brown and Grizzly Bears

AVERAGE LENGTH (including tail): 7 feet (2.1 m)
AVERAGE HEIGHT AT SHOULDER: 3¾ feet (1.1 m)
AVERAGE WEIGHT (coastal): 600 pounds (272 kg)
AVERAGE WEIGHT (grizzly): 350 pounds (160 kg)

In contrast to the black bear's preference for dense forests, brown bears (*Ursus arctos*) inhabit open forests. They inhabit much the same geographic areas as black bears in Alaska, western Canada, and the northwestern United States, the latter in isolated but expanding areas of Idaho, Montana, Washington, and Wyoming.

Grizzly bears are simply brown bears that live inland, as opposed to along the coast. They differ somewhat from coastal brown bears: for example, the grizzly's coat is lighter than that of a coastal brown bear. The name grizzly comes from the frosted, or grizzled, appearance of white-tipped hairs on the bear's shoulders, hump, and back.

Most other differences are attributed not to genetics but to diet. Instead of the rich grasslands and salmon-filled streams inhabited by their coastal cousins, grizzlies occupy inland forests, tundra plains, and high mountain meadows, where food is much less plentiful and their diet consists mainly of plants and berries. As a result of having to work harder to fill their bellies, grizzlies tend to be smaller and more defensive than their coastal cousins.

Grizzlies and other brown bears are active mainly during the morning and early evening. However, while chowing down in preparation for hibernation, they look for vittles throughout the day. As winter approaches and food becomes scarce, the brown bear forages farther afield and may wander into an area where it might not otherwise be seen. In states where grizzlies are a known threat, Defenders of Wildlife (*www.defenders.org*) may help you protect your poultry by designing a bear-resistant electric fence and reimbursing part of the cost.

Brown bears seasonally travel seeking areas of abundant food. Where a male black bear's home range may encompass 10 to 60 square miles, a brown bear typically takes in 115 to 270 square miles. The brown bear has a widely varied diet that includes grass and leaves, nuts and berries, fruit, worms, beetles, ants, termites, and other insects. Brown bears also eat rodents, foxes, and other small animals and sometimes larger animals, like sheep and deer, as well as carrion. In short, their diet is similar to that of a black bear.

One big difference is that brown bears are diggers, while black bears are not. Brown bears turn over rocks to find grubs and dig in the ground for such delicacies as roots, tubers, corms, bulbs, and ground squirrels. As a result, they have a well-developed mass of shoulder muscles that appears as a high point, or hump. The nondigging black bear, by contrast, has no shoulder hump; the high point on its back is above its rump.

This distinction is important, since brown bears are easily confused with black bears. Both can have coat colors in a range of browns. And, although mature brown bears are quite a bit bigger than most black bears, a young or female brown bear can be approximately the size of a mature male black bear. The hump helps you identify a brown bear from a distance.

Other identifying characteristics are the facial profile and the size and shape of the ears.

A black bear's facial profile is flat or rounded (convex) from the forehead to the nose, while a brown bear's facial profile is dished (concave). The black bear's ears are erected and pointed; a brown bear's ears are small and round.

Although you wouldn't want to get close enough to examine a bear's claws, another distinction is that the black bear's claws are dark in color, sharply curved, and less than 2 inches long — more suitable for climbing trees. A brown bear's front claws are pale in color, slightly curved, and 2 to 4 inches long — more suitable for excavating. A heavy brown bear can't climb nearly as well as a black bear. At any rate, if you can see a bear's claws from a distance, it's likely a brown bear.

What's the importance of knowing whether a brown bear or a black bear has designs on your poultry? Well, you can usually repel a black bear by making lots of noise and fuss — yelling at it, waving your arms, throwing rocks, or bonking it on the nose with a flashlight or your fist, if it comes to that (let's hope it doesn't). Try any of those tactics on a brown bear and you're liable to end up on the menu.

If you encounter a brown bear, slowly back away. Do not turn your back and do not run. When you get to a safe place, call your local animal control office, wildlife agent, or sheriff's department and report that a brown bear is hanging around your yard. Let authorities who know what they're doing deal with it.

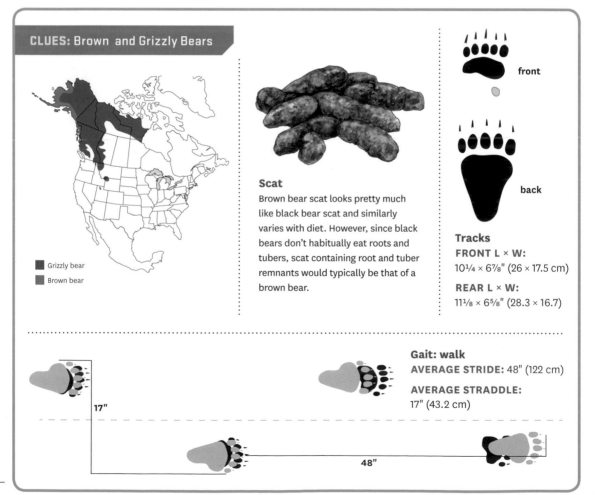

CLUES: Brown and Grizzly Bears

Grizzly bear
Brown bear

Scat
Brown bear scat looks pretty much like black bear scat and similarly varies with diet. However, since black bears don't habitually eat roots and tubers, scat containing root and tuber remnants would typically be that of a brown bear.

front

back

Tracks
FRONT L × W:
10¼ × 6⅞" (26 × 17.5 cm)

REAR L × W:
11⅛ × 6⅝" (28.3 × 16.7)

Gait: walk
AVERAGE STRIDE: 48" (122 cm)

AVERAGE STRADDLE:
17" (43.2 cm)

17"

48"

Bears: A Closer Look

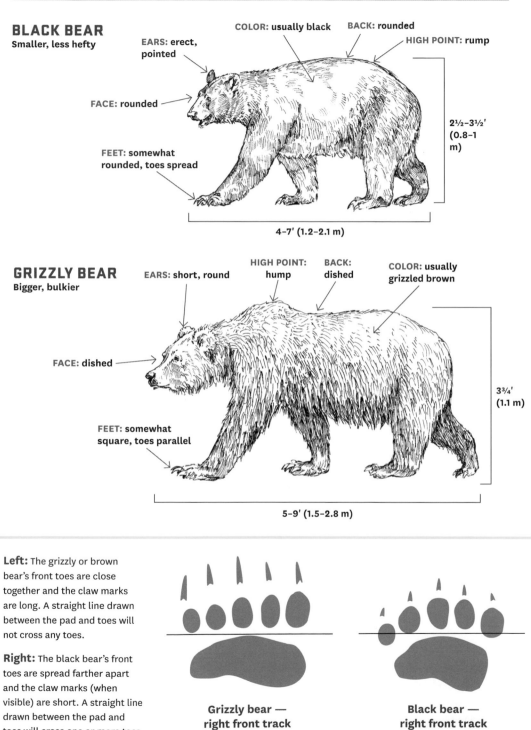

BLACK BEAR
Smaller, less hefty

EARS: erect, pointed

COLOR: usually black

BACK: rounded

HIGH POINT: rump

FACE: rounded

FEET: somewhat rounded, toes spread

2½–3½' (0.8–1 m)

4–7' (1.2–2.1 m)

GRIZZLY BEAR
Bigger, bulkier

EARS: short, round

HIGH POINT: hump

BACK: dished

COLOR: usually grizzled brown

FACE: dished

FEET: somewhat square, toes parallel

3¾' (1.1 m)

5–9' (1.5–2.8 m)

Left: The grizzly or brown bear's front toes are close together and the claw marks are long. A straight line drawn between the pad and toes will not cross any toes.

Right: The black bear's front toes are spread farther apart and the claw marks (when visible) are short. A straight line drawn between the pad and toes will cross one or more toes.

Grizzly bear — right front track

Black bear — right front track

OPOSSUMS

AVERAGE LENGTH (including tail): 2½ feet (76 cm)
AVERAGE HEIGHT: 11 inches (28 cm)
AVERAGE WEIGHT: 10 pounds (4.5 kg)

One of the world's oldest surviving mammal species, opossums (*Didelphis virginiana*) once lived alongside dinosaurs. Today, they are the only marsupials inhabiting North America. The word *marsupial* derives from the Latin *marsupialis,* which means having a pouch and refers to the pocket (or marsupium) attached to the female's abdomen. Marsupial babies are born before they are well enough developed to enter the world. Instead, they continue to develop while living and nursing snug inside the mother's pouch. The well-known, much larger kangaroo of Australia is distantly related to our North American opossum.

TYPICAL SIGNS. Opossums typically forage throughout the night, but they are active during the day when raising young, when food is scarce, or to take advantage of winter's warmer daytime temperatures. Generally a 'possum will scrounge close to its den, but it may travel a couple of miles if necessary to find enough to eat. Since it does not hibernate, does not cache food, and has little body fat, it spends most of its waking hours on the alert for things to eat.

'Possums are not fussy eaters, although they prefer meat to plants. They will eat whatever they run across, be it an egg or the hen that laid it. They have an acute sense of smell that may well lead them to a henhouse, and a terrific memory for food sources. Once an opossum finds its way into a chicken coop, it may consider it to be a private pantry offering a continuous supply of chicken feed, eggs, and fresh meat. While the pickings are easy, the 'possum may den in a barn loft or a crawl space under the chicken coop.

The 'possum is a messy eater. When it chews up eggs, it leaves smeared contents and bits of shell in the nest. A young animal may then nap in the nest, where you'll find it in the morning, curled up snug. A 'possum will eat young poultry almost entirely, perhaps leaving only a few matted feathers. Among older poultry, the 'possum will target a chicken sleeping on the roost. It may start on the leg, breast, abdomen, or cloaca and generally favors innards over muscles. It will eat until it's had enough and leave behind whatever remains.

If you catch a 'possum in the act, it will typically waddle away, crawl into a hiding place, or climb a tree. If it feels trapped, it may turn and bare its teeth, or it may fall over dead.

Not really. It just looks dead, lying motionless on its side with its eyes open and its tongue lolling out. Sometimes, for good measure, it will drool, urinate, defecate, and release a foul-smelling anal fluid. Prod the animal and it won't react. It may remain thus for a few minutes or several hours, until it perceives that danger has passed. Whether this charade, called playing possum, is a deliberate ruse or an involuntary response to fear remains a puzzle.

APPEARANCE. The 'possum has a long pointed snout, pink nose, round hairless ears, small dark eyes, and lots of sharp teeth — 50, to be exact — designed for crushing and grinding. A cornered opossum opens its mouth wide and growls or hisses while displaying its formidable teeth to frighten intruders.

It's mostly bluff, though. Slow and cautious, 'possums aren't good fighters and normally won't bite or attack. They'd rather flee than fight. But a 'possum can't run fast on its short legs, and if it or its offspring are truly threatened it won't hesitate to use those fearsome teeth in defense.

An opossum is about the size of a house cat but with a thicker body and shorter legs. Males are slightly bigger than females and have larger, fiercer-looking canine teeth. Size varies geographically, with larger specimens found in colder climates. Within a given area, size also depends on the abundance of food. The weight of an adult generally ranges between 5 and 15 pounds, the shoulder height is 8 to 14 inches, and the length is 2 to 3 feet, including the tail.

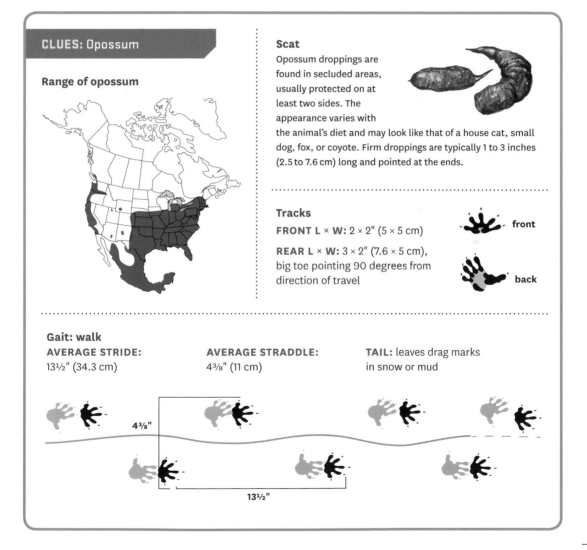

CLUES: Opossum

Range of opossum

Scat

Opossum droppings are found in secluded areas, usually protected on at least two sides. The appearance varies with the animal's diet and may look like that of a house cat, small dog, fox, or coyote. Firm droppings are typically 1 to 3 inches (2.5 to 7.6 cm) long and pointed at the ends.

Tracks

FRONT L × W: 2 × 2" (5 × 5 cm)

REAR L × W: 3 × 2" (7.6 × 5 cm), big toe pointing 90 degrees from direction of travel

front

back

Gait: walk

AVERAGE STRIDE: 13½" (34.3 cm)

AVERAGE STRADDLE: 4⅜" (11 cm)

TAIL: leaves drag marks in snow or mud

4⅜"

13½"

HABITAT AND RANGE. The opossum is basically a southern species that has expanded its range westward to Colorado and as far north as southern Ontario (and occasionally Quebec). Its range is limited by its ability to withstand cold weather; in the more northerly areas, opossums typically lose ear tips and tail ends to frostbite. Isolated areas of the Southwest and nearly the entire length of the West Coast are now also inhabited by opossums, offspring of captives that either were released or escaped.

BEST DETERRENTS. A 'possum has nimble fingers for opening simple fasteners, latches, and containers. It is an excellent climber, ascending a fence hand-over-hand and using its prehensile tail for grip and balance. And it can squeeze through small spaces. Keeping a 'possum away from poultry can be challenging.

For starters, do not leave poultry feed out overnight. Keep the poultry yard clear of dense bushes, piles of old lumber, and other hiding places. Lock poultry indoors overnight and secure potential coop access points with hardware cloth.

A young opossum can slip through a wire fence with mesh size larger than 3 inches. Any 'possum can easily climb a wood or wire fence. An electrified scare wire, 6 inches or so above the ground, will deter climbing. A tree proximate to the fence may be climbed to gain access by dropping down into the yard from an overhanging branch. Since 'possums don't jump, 18 inches of sheet metal wrapped around the base of the tree will prevent climbing.

If the coop is elevated less than 2 feet off the ground, block underneath access with hardware cloth or concrete blocks to discourage denning. If a 'possum already has a den under the coop, make it uncomfortable with flashing lights and talk radio. After the 'possum exits at dusk, locate the nest and tear it apart. Once the 'possum, and any babies, have vacated, securely seal access.

Aside from securing your poultry, you might think twice about discouraging the presence of opossums. They are magnets for ticks. When a 'possum isn't sleeping or eating, it's grooming, like a cat, and in doing so swallows tons of ticks — 5,000 each season, at one estimate. 'Possums eat snakes. Not just egg-eating snakes, but also venomous snakes. 'Possums serve as a cleanup crew, ridding the area of food items that attract rats and mice. And if a rodent does come along, the 'possum will eat that, too. But if you just can't tolerate having a 'possum wandering around your yard, wait a couple of days and it will move on.

STATE ANIMAL

You would think the Virginia opossum (*Didelphis virginiana*) would be a state animal of Virginia, but no — it's North Carolina's official state marsupial.

No Shortage of 'Possums

The first time I saw an opossum on our Tennessee farm, I had gone out to the garden to toss some kitchen scraps into the compost bin and startled a 'possum snacking in the bin. Or, rather, I should say the 'possum startled me with its awesome display of teeth. Not knowing yet that 'possums are nonaggressive, I backed away, duly impressed.

I soon learned that Tennessee has no shortage of opossums, which means no shortage of opossum/poultry incidents. More than once, we have gone into the chicken coop to collect the morning eggs only to find a young 'possum taking a snooze in a nest. Since it doesn't happen often, we find it more amusing than annoying.

However, what *is* annoying is when an opossum takes a bite out of a living chicken. One time, a 'possum got into a temporary chicken shelter and bit chunks out of the breasts of several young New Hampshires. Another time, a 'possum got into our Silkie house and gnawed on one of the little hens. And that Silkie house was surrounded by chain-link fence with an electrified scare wire. I chased the 'possum, but before it could show me how it got inside the fence, it keeled over and played dead. Next morning, it was gone.

Opossums are solitary foragers, so usually we encounter only one at a time. However, every rule has an exception. One time, three 'possums enjoyed feasting in a coop full of growing fryers. From their

relative sizes, they appeared to be a mama with two almost grown youngsters.

These various incidents were far between, having occurred over a span of 36 years. As annoying as they are, the comfort is that opossums are nomadic and nonterritorial, and they don't hang around one place for more than a few days.

We might spy a wandering 'possum, or catch one on our trail cam, for several days in a row and then not see another for weeks on end. Meanwhile, when one does manage to find its way into one of our poultry yards, we try to figure out how it got in and take measures to ensure it won't happen again.

SKUNKS

Striped Skunk · Spotted Skunks · Hooded Skunk · Hog-Nosed Skunk

Striped Skunk

AVERAGE LENGTH: 2 feet (30 cm)
AVERAGE HEIGHT: 8 inches (20 cm)
AVERAGE WEIGHT: 8 pounds (3.6 kg)

Western striped skunk

Striped skunks (*Mephitis mephitis*) are best known for their offensive smell. Indeed, the taxonomic name *mephitis* means noxious vapor in Latin, and in the case of the striped skunk is doubled up. The "noxious vapor" issues from two anal stink glands that are larger and more powerful than those of a weasel. Each gland has a little nozzle surrounded by muscles that contract to shoot either a fine mist or a hard stream that can hit a target as far away as 10 feet or even, with less accuracy, 15 feet. The sulfuric yellow oily fluid has a nauseating odor that may be detected up to a mile away. Sprayed into the eyes, it causes temporary blindness.

Luckily, skunks don't go around shooting at random. In fact, they are reluctant to fire at all because they have only a limited supply of ammunition. Once depleted, it takes about a week to regenerate, meanwhile leaving the skunk basically defenseless. A skunk therefore much prefers to retreat from danger and is likely to spray only when it has been startled, cornered, or harmed, is defending young, or is itself young and inexperienced.

Typically before a striped skunk sprays, it will give fair warning. It will rapidly drum on the ground with its front feet, arch its back, lift and puff out its tail, issue threatening noises (chatter, snarl, spit, growl, or hiss), and make brief forward lunges. If the object of its displeasure does not heed these warnings and move away, the skunk will turn tail and make short backward scoots while letting loose. Pee-ew!

The stench may linger for months, especially where spray has penetrated something permeable, like wood. A skunk once got into one of the stalls in our barn. It was apparently young and easily frightened, because when I moved to open the nearest door so it could get out, it fired a bit too readily. Luckily the spray didn't land on me, but unluckily it hit the barn wall and stank up the place for several weeks. The next time I saw a skunk in the barn I let it be; eventually it wandered off on its own.

Despite their odorous nature, skunks perform a valuable service in their choice of foods. They prefer to dine on grubs and any kind of insect (including cockroaches) but will eat just about anything, including poisonous snakes, scorpions, black widow spiders, and wasps. They also eat a variety of plant matter, as well as rodents, carrion, and garbage. But given a chance, they also enjoy snacking on eggs, baby poultry, and sometimes even a mother hen.

TYPICAL SIGNS. Skunks typically forage at dawn and dusk but may look for food at any time of day or night. A striped skunk doesn't climb, but it sure can dig. It will dig under a fence and feed on any eggs left in nests overnight, opening shells at one end and crushing the broken edges inward with its nose in an effort to lick out the contents.

A skunk will occasionally kill one or two chickens or ducks, eating only the head and neck or sometimes the entrails, but not the muscle. A striped skunk doesn't move fast enough to catch poultry during the day but will target a bird that is either on a nest or roosting at night. It also relishes unborn eggs and hatchlings.

If a dense shrub or other object is near the outside of the poultry fence, a striped skunk may use it to get over the fence and into the yard. Since it won't be able to climb back out, you may find it trapped inside the yard and wonder how it got there. A freshly dug hole under a coop or other building may indicate that a skunk is denning for a night or three — unless it is a female looking for a place to give birth, in which case she plans to stay longer. Look for 2-inch black or white hairs clinging to wood or other rough surfaces around the opening.

A striped skunk's tracks are similar in size and shape to those of a house cat, except the skunk has five toes on all four paws, instead of four, and its long nonretractable claws are more likely to make marks. Also the skunk usually leaves staggered tracks, unlike a cat whose back foot steps on the front track.

APPEARANCE. A striking black-and-white coat makes the striped skunk one of the most readily recognizable North American mammals. The typical pattern is a jet-black

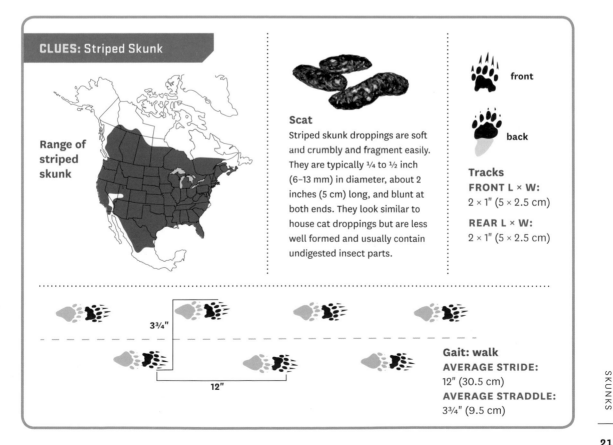

CLUES: Striped Skunk

Range of striped skunk

Scat

Striped skunk droppings are soft and crumbly and fragment easily. They are typically ¼ to ½ inch (6–13 mm) in diameter, about 2 inches (5 cm) long, and blunt at both ends. They look similar to house cat droppings but are less well formed and usually contain undigested insect parts.

front

back

Tracks
FRONT L × W:
2 × 1" (5 × 2.5 cm)

REAR L × W:
2 × 1" (5 × 2.5 cm)

3¾"

12"

Gait: walk
AVERAGE STRIDE:
12" (30.5 cm)
AVERAGE STRADDLE:
3¾" (9.5 cm)

coat with a white stripe that starts on the head and splits at the shoulders into two stripes that extend to a bushy black-and-white tail. The stripes on individual skunks vary in length and width so much that some skunks appear nearly all black while others look completely white.

The striped skunk is about the size of a house cat, but with short, stocky legs and relatively large feet. Weights range from about 3 to 12 pounds, with an average of about 8 pounds. Including a 7- to 10-inch tail, total length is about 2 feet, and the height at the shoulders averages 8 inches. Males are slightly larger than females, and northern dwellers are larger than those living farther south.

The striped skunk is wide at the rear and narrow toward the front, with a proportionally small head, pointed nose, and small ears and eyes. Its senses of smell, hearing, and sight are so poor that a skunk intent on foraging may walk right past you. Once I was walking in the woods behind my house when I stopped to watch a skunk snuffling through the leaves. It passed within inches of my feet without noticing I was there.

Like the skunk I encountered that day, striped skunks are typically easygoing and in little hurry as they meander along with short, shuffling steps. They walk flat-footed, like a raccoon, and like a raccoon they have an acute sense of touch.

HABITAT AND RANGE. Striped skunks are the most common skunks in North America. They make themselves at home throughout the southern half of Canada and most of the contiguous states. They adapt to a wide range of habitats including brushy woodlands, agricultural areas, suburbs, and even urban areas. They are less fond of dense mountain forests and they stay away from arid Mojave and Colorado deserts, preferring to remain within 2 miles of a source of drinking water.

BEST DETERRENTS. Since striped skunks can't climb a slope having more than a 45-degree angle, they are easily deterred by a well-built fence. Remove any bushes or other objects near the outside of the fence that might allow a skunk to clamber up and over. Either bury the bottom of the fence and bend it outward into a skirt or lay large, flat paving stones around the outside perimeter to discourage digging.

Prevent skunks from denning beneath the coop by closing access with hardware cloth or some other sturdy barrier. If a skunk has already taken up residence, you might encourage it to move along by lighting the den area with flashing lights and leaving a radio tuned to a talk station. Seal off access only after you are certain the skunk has left.

SKUNK ODOR REMOVAL

Skunk spray on a wall, clothing, pet, or poultry may be neutralized with the following solution:

- 1 quart (0.95 L) 3% hydrogen peroxide
- ¼ cup (45 g) baking soda
- 1 teaspoon (5 ml) liquid dish detergent

Wearing disposable gloves, immediately apply the solution as a rub or shampoo, concentrating on areas where you can see the yellow musk. Let it soak in for five minutes, then rinse it away with plain water. Repeat as necessary. A fan used to increase ventilation will hasten odor dissipation. Eyes that have been sprayed should be flushed with plain water as soon as possible.

The Egg-Rolling Skunk

Behind our house is a small pond on which we once kept a pair of Embden geese. The goose laid her eggs in a small doghouse-like structure. Mysteriously, the eggs disappeared almost as fast as they were laid.

The goose house was situated just outside our bedroom window. One night my husband and I were awakened by a loud ruckus raised by the pair of geese. We got to the window in time to see a small striped skunk rolling a big goose egg down the slope toward the nearby woods line.

Gander and Mrs. Gander stood side-by-side stretching their necks and honking, but the persistent skunk got away with its prize. I admit we had to laugh at the effort the little skunk made to move that big egg. It looked so earnest. And so adorable. And now we knew where the goose's eggs were going.

In the morning, we were less amused when we discovered Gander had a sizable bite taken out of his chest. So it was off to the vet, who explained that Gander, not being a mammal, most likely would not get rabies from the skunk bite. But the wound might become infected, so we were sent home with antibiotic pills and ointment to administer to Gander daily.

Meanwhile, having received no more visits from the skunk, Mrs. Gander managed to hatch out a single gosling. While the little guy grew, Gander stood guard over him as if he realized how lucky the precious gosling was to have avoided becoming dinner for a neighborhood skunk.

More Skunks

The skunk family (Mephitidae) in North America includes five species, of which the striped skunk is the largest and most common. Nearly as common are two species of spotted skunk, and least common are hog-nosed and hooded skunks. The word *skunk* derives from the Algonquian name for the animal, *seganku*. The word *Mephitidae* is Latin for bad smell, one of the attributes mephitids have in common.

Other traits they share include the following:

- They rarely spray without good reason and usually first give a warning.

- They typically wear distinguishable bold black-and-white coats.

- They have long, bushy tails.

- They have short legs, large feet, and long, sharp claws.

- Digging is their main technique for finding food.

- They eat whatever they can find or catch.

- They are more likely to pilfer eggs than attack live poultry.

- They do not store food but will raid the caches of weasels and others.

- They are primarily crepuscular but may forage at any time of day or night.

- They are typically nonaggressive.

- They can make a variety of sounds but normally remain quiet.

- They are sometimes erroneously called polecats or civet cats.

- Signs and deterrents are similar for all skunk species.

Skunks: A Closer Look

EASTERN SPOTTED
WEIGHT: 2 lbs (0.9 kg)

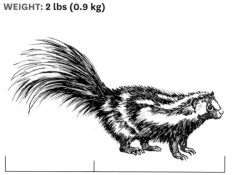

TOTAL LENGTH: 20" (51 cm)
TAIL LENGTH: 40% of total length

WESTERN SPOTTED
WEIGHT: 1 lb (0.5 kg)

TOTAL LENGTH: 15" (38 cm)
TAIL LENGTH: 40% of total length

STRIPED
WEIGHT: 8 lbs (3.6 kg)

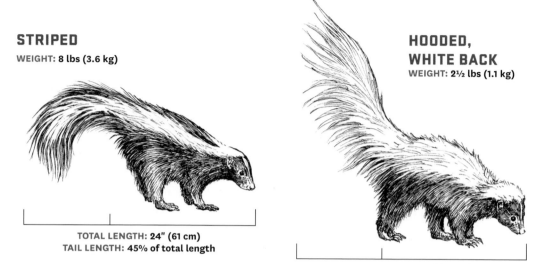

TOTAL LENGTH: 24" (61 cm)
TAIL LENGTH: 45% of total length

HOODED, WHITE BACK
WEIGHT: 2½ lbs (1.1 kg)

TOTAL LENGTH: 26" (66 cm)
TAIL LENGTH: 50% of total length

HOG-NOSED
WEIGHT: 4 lbs (1.8 kg)

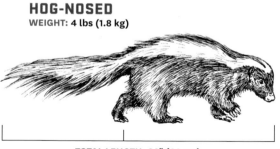

TOTAL LENGTH: 26" (66 cm)
TAIL LENGTH: 30% of total length

HOODED, BLACK BACK
WEIGHT: 2½ lbs (1.1 kg)

TOTAL LENGTH: 26" (66 cm)
TAIL LENGTH: 50% of total length

Eastern spotted skunk

Western spotted skunk

Spotted Skunks

EASTERN SPOTTED SKUNK

AVERAGE LENGTH (including tail):
20 inches (51 cm)
AVERAGE HEIGHT AT SHOULDER: 5 inches (13 cm)
AVERAGE WEIGHT: 2 pounds (900 g)

WESTERN SPOTTED SKUNK

AVERAGE LENGTH (including tail):
15 inches (38 cm)
AVERAGE HEIGHT AT SHOULDER: 4 inches (10 cm)
AVERAGE WEIGHT: 1 pound (450 g)

Spotted skunks (*Spilogale* spp.) are the smallest members of the skunk family, the most weasel-like in appearance, and the most agile. They are swift enough to handily catch fast rodents and other small mammals, and small enough to pursue prey into its tunnel. They can also more easily catch fleeing poultry and therefore can inflict more damage to a flock than a striped skunk can. Spotted skunks are also the only skunks that can climb with ease, including over fences, and they may den in barn lofts, attics, and hollow trees, as well as closer to the ground.

Although the spotted skunk's spray is more pungent than that of a striped skunk, the animal is less likely to spray. When startled, the spotted skunk prefers to escape danger by climbing the nearest tree or fence post. If it feels threatened enough to spray, the spotted skunk first issues a warning by doing a rapid series of front handstands in an attempt to make itself look bigger and fiercer. To a human, it looks simply comical, at least until the skunk arches its back and discharges its spray forward over its head.

Spotted skunks are of two distinct species: the eastern spotted skunk and the western spotted skunk. The two differ in coat pattern, size, reproduction, choices of habitat, and geographic range — although together they cover most of the United States and share a narrow area where their ranges meet. Both species may be readily identified as spotted skunks by their white spots and white stripes interrupted by black fur. Each individual has a unique set of markings. Tracks are similar to those of a striped skunk, but smaller. Scat, too, is similar to a striped skunk's, but about half the size.

Deterring spotted skunks is basically the same as for striped skunks, with the addition of preventing spotted skunks from climbing into the poultry yard. An electrified scare wire 5 inches above the ground on the outside of the fence will discourage fence climbing. Extension arms at the top of the fence will prevent skunks from being able to get over the top.

Range of spotted skunk

■ Western
■ Eastern

Hooded Skunk

AVERAGE LENGTH (including tail):
26 inches (66 cm)

AVERAGE HEIGHT AT SHOULDER:
7½ inches (19 cm)

AVERAGE WEIGHT: 2½ pounds (1.1 kg)

Hooded skunks (*Mephitis macroura*) live mostly along streambeds and washes in scrubby grasslands or rocky areas of southeastern Arizona, southwestern New Mexico, and western Texas. They are primarily solitary but may congregate to dine at garbage dumps. Unlike other skunks, they do little digging and mainly snack on grasshoppers and other insects, small rodents, and cactus fruits. They tend to gravitate to areas inhabited by humans, where they find ready-made hiding places, as well as plenty of delicious garbage and, when lucky, poultry eggs.

Hog-Nosed Skunk

AVERAGE LENGTH (including tail):
26 inches (66 cm)

AVERAGE HEIGHT AT SHOULDER:
10 inches (25 cm)

AVERAGE WEIGHT: 4 pounds (1.8 kg)

Among the many common designations for a hog-nosed skunk (*Conepatus leuconotus*) are rooter skunk and badger skunk, after the species' propensity to dig up soil looking for bugs and grubs. Its awesome rooting ability is thanks to its strong front legs, long front claws, and a long, wide, hairless snout. One of these skunks can rototill an area 40 feet (12 m) in diameter, even overturning rocks, and leave it looking as if it's been plowed up by feral pigs. Indeed, its biggest competitors for food are feral hogs.

Although some states designate hog-nosed skunks as predators, they are the most insectivorous and least predacious of the skunk species. They enjoy a much less diverse diet than other skunks, eating mainly insects, but sometimes also small rodents, reptiles, fruits, and nuts. Although they tend to avoid encounters with humans, they may venture into a poultry yard to dine on tender young chicken.

Range of hooded skunk

Range of hog-nosed skunk

SKUNKS

RODENTS

Ground Squirrels · Rats · Mice

Ground Squirrels

LENGTH: 13 to 20 inches (33–43 cm); varies by species
WEIGHT: 12 to 32 ounces (0.3–0.9 kg); varies by species

Ground squirrels are a varied group of rodents in the Marmotini tribe of the squirrel family (Sciuridae) that live in underground burrows rather than in trees. The term *ground squirrel* is sometimes used collectively for all members of the tribe. True ground squirrels are midway in size between their smaller cousins, the chipmunks, and their larger cousins, prairie dogs and marmots (including groundhogs).

Classification of North America's 58 currently recognized ground squirrel species is in a state of flux (read "confusion") thanks to the wonders of DNA sequencing. All are herbivores and many rarely, if ever, eat eggs or birds. Those most likely to pester poultry are the larger species: California ground squirrels (*Otospermophilus beecheyi*), Columbian ground squirrels (*Urocitellus columbianus*), Franklin's ground squirrels (*Poliocitellus franklinii*), and rock squirrels (*Otospermophilus variegatus*). They have these features in common:

- They have short legs and strong claws.

- They live in underground burrows.

- They prefer habitats with loose soil for burrow excavation.

- They hole up in their burrows during cold or hot weather.

- They tend to be gregarious and may live in multigenerational colonies.

- They carry food in large cheek pouches to eat later.

- They hoard large amounts of food in burrow caches.

- They stand upright when they want to get a better view.

- They warn family members of danger by peeping or whistling.

- None may be found in the eastern one-third of North America.

Ground squirrels live only in central and western United States and Canada. In some states, some species enjoy varying degrees of protective status, while others are categorized as pests.

TYPICAL SIGNS. Ground squirrels, like tree squirrels, are basically vegetarians that are attracted to poultry yards where grain or other feed is readily available. However, unless tree squirrels have trouble finding other things to eat, they rarely eat eggs or birds. Further, tree squirrels do not hibernate and therefore actively forage year-round.

Most ground squirrels, on the other hand, sleep in their burrows for much of the year. Prior to winter hibernation, when they need extra calories for fattening up, they may become omnivorous and eat insects, caterpillars, small mammals and birds, and eggs.

California ground squirrel

During this time, some ground squirrels may boldly enter a coop to steal poultry eggs. This behavior is especially likely where a lot of ground squirrels compete for limited resources, and one of them discovers the poultry yard as a reliably well-stocked pantry.

One egg is usually enough to satisfy a squirrel's immediate appetite. A ground squirrel sometimes eats the shell and all. A California ground squirrel may leave an empty shell with a hole in the side, or a broken shell, or no shell at all — just a little damp egg residue in the chicken's nest.

A duck egg, with its tougher shell, is not as easy to open as a chicken egg. Franklin's ground squirrels have developed a duck-egg-eating technique that has been well documented on wild duck nesting grounds. The squirrel straddles an egg longwise and either bites into one end, rolls its body against the egg until the shell breaks, or bounces on the egg. Clues of duck egg predation include eggs missing over a course of time, partially eaten eggs, or shells with large irregular end holes surrounded by finely serrated edges. When shells are found, they would typically be scattered near the nest or trailing away from the nest.

Although most ground squirrels do not have natural predatory skills, some species occasionally dine on rabbits, rodents (including other ground squirrels), small wild turkeys, and easy-to-catch young poultry. Rock squirrels, in particular, are fond of flesh. Franklin's ground squirrels sometimes eat duck hatchlings or growing ducks. A typical sign of ground squirrel predation is a dead duckling with the organs consumed but the head intact. California ground squirrels will occasionally attack chicks or partially grown chickens, typically biting off the feet and beak and chewing soft flesh.

Since ground squirrels are diurnal, they are easier to catch in the act than predators that prowl under cover of darkness. The usual time of activity varies with species and climate. California ground squirrels, for instance, are typically active from midmorning through late afternoon, while Columbian and rock squirrels are most active in the early morning and late afternoon, thus avoiding midday heat.

BE SAFE!

Ever since rat-infested steamships introduced bubonic plague into the United States in 1900, the disease has spread from rats to other rodent species. Ground squirrels are susceptible, and entire colonies have been wiped out. Bubonic plague particularly can be a problem where large numbers of ground squirrels live in close proximity. Seeing an unusual number of squirrels, or any rodents, dead for no apparent reason is cause for concern and should be reported to local public health officials.

Do not handle dead rodents. Bubonic plague can be transmitted to a human by fleas from infected squirrels. Without prompt medical attention, this serious illness may lead to complications or death.

BEST DETERRENTS. Keeping ground squirrels out of the poultry yard presents a unique challenge. For starters, they are diurnal, so closing up poultry at night is not helpful. Their extensive, multiple-entrance burrow systems render most anti-digging measures rather ineffective. Further, ground squirrels are pretty good climbers and may go over a fence as well as under it. The best deterrent is a secure coop, with a concrete floor, and well-attached hardware cloth lining the bottom and top of the run.

To evade their many predators, ground squirrels typically forage close to a burrow entrance where they can quickly dive to safety. Measures that prevent burrowing near the coop offer a big step toward preventing predation. Make the area less attractive by clearing out brush, wood piles, and other debris that could provide cover for burrow entrances.

A guardian dog may be helpful, as may a feral cat. Celeste Tittle once had a problem with ground squirrels at Ham and Eggs Ranch in Norco, California, where she raises exotic poultry. Nine years ago someone dumped two pregnant feral cats in front of her ranch and she hasn't had a squirrel problem since. Celeste doesn't pet the multiple cats or feed them but leaves them alone to do their job. The cats don't bother her mature poultry, although she does protect baby chicks in pens covered with half-inch hardware cloth.

CLUES: Ground Squirrels

Range of ground squirrels

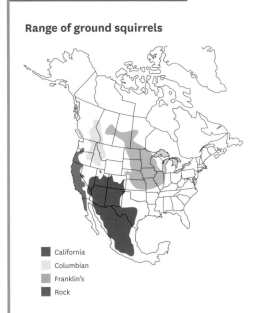

- California
- Columbian
- Franklin's
- Rock

Scat
The size and shape of ground squirrel scat vary with species and diet. Typically it is tubular, rounded at one end and tapered at the other, ³/₁₆ inch in diameter (5 mm), ½ inch (12 mm) long, and often found near a burrow entrance.

Tracks
FRONT L × W: 1¼ x 1" (3.2 x 2.5 cm)

REAR L × W: 1³/₈ x 1¹/₈" (3.5 x 2.9 cm)

front

back

Gait: walk
STRIDE: 9½" (24.2 cm)
STRADDLE: 2⅞" (7.3 cm)

Foiling Ground Squirrels

Diana Mitchell, California, chicken keeper

For a while I had a big problem with California ground squirrels. They were eating everything I put out for my girls. I allowed the hens to free-range the yard, so the main door was wide open and the squirrels were going right into the coop to eat the food. Later on we added an extension to our small coop to create a larger, completely enclosed area. However, I still let them free-range, just less often. Having a dog was also a helpful deterrent against ground squirrels.

The trees running along our back fence were another issue. They were tall enough to reach the power wires. The ground squirrels — along with opossums, rats, and mice — were using the trees, power poles, and power lines to get from one location to another. The trees had little berries the critters were eating, which, along with the free chicken food, made our yard a great place to hang out.

The trees had some structural issues and, with the critter problem, it made sense to trim the trees well below the power lines. We later reluctantly removed the trees, as they were diseased. Both the trimming and removing helped reduce the number of critters frequenting our yard. Once they didn't have easy access, wire to tree, they moved into someone else's yard.

More recently we added an open-top run along the back fence where the trees used to be. The ladies are in there now, more than free-ranging the yard, which is beneficial because sometimes they dig up things we don't necessarily want them digging up.

When we added the enclosed run, I made a money decision to save a few dollars by getting wire with bigger holes. It cost me more in the end because the local birds can fit right through the wire. Now I'm no longer feeding ground squirrels and other four-legged critters, but I'm feeding all the local birds. I do like having wild birds in the yard, so it's not the worst thing that could happen. At least now the ground squirrels and other critters are looking in someone else's backyard and are no longer my problem.

Ground Squirrels: A Closer Look

CALIFORNIA GROUND SQUIRREL

WEIGHT: 26 oz (7.4 kg)

HABITAT: fencerow

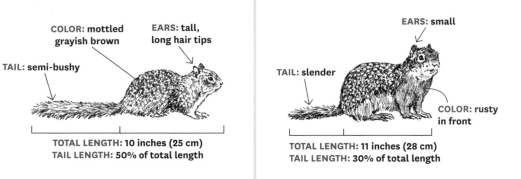

COLOR: mottled grayish brown

EARS: tall, long hair tips

TAIL: semi-bushy

TOTAL LENGTH: 10 inches (25 cm)
TAIL LENGTH: 50% of total length

COLUMBIAN GROUND SQUIRREL

WEIGHT: 16 oz (4.5 kg)

HABITAT: brush

EARS: small

TAIL: slender

COLOR: rusty in front

TOTAL LENGTH: 11 inches (28 cm)
TAIL LENGTH: 30% of total length

FRANKLIN'S GROUND SQUIRREL

WEIGHT: 22 oz (6.2 kg)

HABITAT: tall, dense grasses

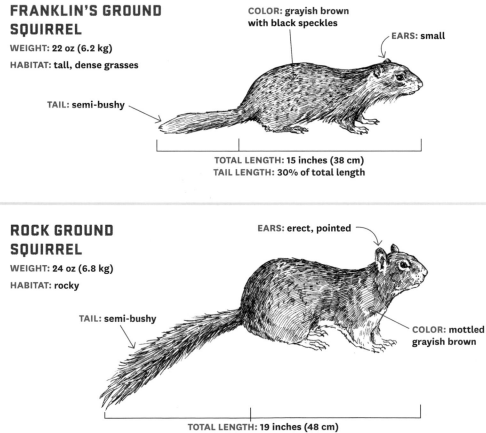

COLOR: grayish brown with black speckles

EARS: small

TAIL: semi-bushy

TOTAL LENGTH: 15 inches (38 cm)
TAIL LENGTH: 30% of total length

ROCK GROUND SQUIRREL

WEIGHT: 24 oz (6.8 kg)

HABITAT: rocky

EARS: erect, pointed

TAIL: semi-bushy

COLOR: mottled grayish brown

TOTAL LENGTH: 19 inches (48 cm)
TAIL LENGTH: 40% of total length

Rats

AVERAGE LENGTH (including tail): 15 inches (38 cm)
AVERAGE WEIGHT: 9–14 ounces (255–397 g)

Two species of rat (*Rattus* spp.) commonly inhabit the United States and Canada — Norway rats (*R. norvegicus*) and roof rats (*R. rattus*). These species are rodents in the Murinae subfamily of the Muridae family and have the following features in common:

- They have long tails and short legs.

- They have long, strong nails for digging.

- They have continually growing chisel-like incisors kept worn down through gnawing.

- They have poor eyesight and are colorblind.

- Their senses of smell, taste, touch, and hearing are excellent.

- They are affectionate and playful.

- They are intelligent and have outstanding memories.

- They forage within a home-range radius of 98 to 164 feet.

- They are cautious of anything new in their environment.

- They can breed at two to three months of age.

- They have a life span of about a year, during which each female weans some 20 young.

- They must drink ½ to 1 ounce of water daily.

- They are omnivorous and include meat in their diet.

Rats are attracted to poultry yards by feed in troughs or spilled on the ground, the ready availability of water, and protection from the elements. They can eat an amazing amount of feed, as well as spread diseases on their feet and through urine, droppings, hair, and fleas left in feed troughs or storage bins.

TYPICAL SIGNS. Once attracted to a coop, rats will eat eggs in nests, leaving an empty shell with a chipped-away hole in the side or end. Rats will also snack on baby chicks or on the toes of sleeping chickens. A flock that anticipates being plagued by rats during the night becomes particularly restless as dusk approaches.

Since rats are nocturnal, you are less likely to spot furry creatures scurrying around than to see signs: holes gnawed through walls, tunnel openings (in soil, in litter, or under floors), and droppings around stored feed. The more droppings you see, the worse your infestation. If you happen to see a rat in the daytime, the situation is out of control — the rat population has grown so large that dominant individuals are forcing the weaker ones to feed in daylight.

Rats generally live outdoors during warm months, moving to indoor comfort during late fall or early winter as cold weather approaches. Their tunnels, and the holes they gnaw through walls, provide access for other predators.

STAY SAFE!

Rodents can carry a number of serious diseases that affect humans, some of which also affect poultry. Rodents can also transmit a variety of mites, lice, fleas, and ticks to poultry. The Centers for Disease Control and Prevention website (www.cdc.gov/rodents/index.html) offers excellent tips on how to safely clean up a rodent-infested area.

RODENTS

Norway rat

Roof rat

NORWAY RATS (*Rattus norvegicus*) are so named because of a mistaken belief that they originated in Norway, which was disproved with the discovery that they inhabited other countries before migrating to Norway. They go by many other names including barn rat, basement rat, brown rat, common rat, gray rat, sewer rat, street rat, water rat, and wharf rat.

Although Norway rats are the largest rodents that commonly live among humans, they are rarely detected by tracks. They usually travel alongside a wall, developing smooth trails used by multiple rats that obliterate each other's tracks. As a sign, look instead for clusters of up to 20 oval droppings, usually dark brown or black and of uniform consistency, ¾ inch long and ¼ inch in diameter.

ROOF RATS (*Rattus rattus*) are so called because they typically live in a building's upper level, or in trees. They are also known as black rats, fruit rats, house rats, ship rats, or tree rats. Unlike less nimble Norway rats, roof rats are agile climbers and often travel along pipes, utility lines, and fence tops.

A roof rat is more slender than a Norway rat, reaching a maximum weight of about three-quarters of a pound. Although its total length is similar to that of a Norway rat, the roof rat has a longer tail in relation to the length of its body, as well as bigger eyes and ears and a more pointed snout. Its coat is smooth and usually either gray or black.

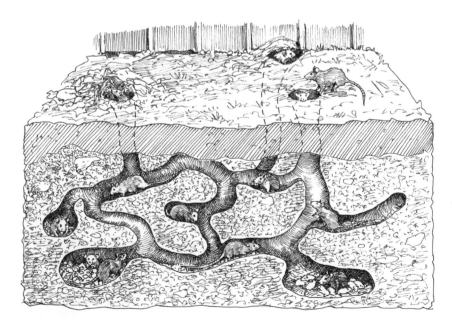

◄ Norway rat tunnel

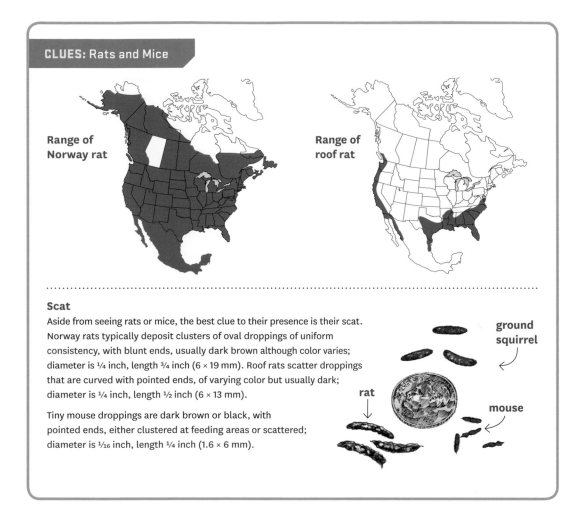

Range of
Norway rat

Range of
roof rat

Scat

Aside from seeing rats or mice, the best clue to their presence is their scat. Norway rats typically deposit clusters of oval droppings of uniform consistency, with blunt ends, usually dark brown although color varies; diameter is ¼ inch, length ¾ inch (6 × 19 mm). Roof rats scatter droppings that are curved with pointed ends, of varying color but usually dark; diameter is ¼ inch, length ½ inch (6 × 13 mm).

Tiny mouse droppings are dark brown or black, with pointed ends, either clustered at feeding areas or scattered; diameter is ¹⁄₁₆ inch, length ¼ inch (1.6 × 6 mm).

ground
squirrel

rat

mouse

Mice

Although mice (*Mus* spp.) look like rats, they are much smaller, don't break and eat poultry eggs (although if a mouse finds an egg already broken, you can bet it will take advantage), and don't attack chickens. Rather, the other way around — a chicken will eat any mouse it can catch. Rats, too, eat mice, so if you have a rat problem you probably don't have mice, or at least not for long.

Besides potentially spreading diseases and parasites, a mouse can consume about 1 pound of chicken feed in a year. When you figure how

rapidly mice reproduce — each female weans 30 to 35 young per year, and the offspring are ready to mate by two months of age — it's not hard to understand how fast chicken feed can disappear when mice get out of control.

Mice are most active at dusk and dawn, but they feed at any time of day. They are nimble jumpers and climbers and can fit through an opening as small as ¼ inch. They are also good swimmers, although if one falls or jumps into a bucket of water from which it can't climb out, it will tread until it drowns from exhaustion.

Rat and Mouse Deterrents

By far the best way to discourage rats and mice is to make the poultry area unattractive to them. Rodents avoid exposure to predation by shunning open spaces, so begin by reducing weeds and keeping grass and shrubs trimmed. Clear away loose lumber, debris piles, and other clutter that can conceal tunnel entrances.

Eliminate the attraction of feed by removing outdoor feeders each night, or filling them in the morning with only as much as your flock will consume before nightfall. To discourage mice from feeding during the day, use either hanging feeders that are difficult for them to reach or treadle feeders, which mice are not hefty enough to operate.

Remove feed from sacks, which are easy to gnaw through, and store it in a galvanized can with a tight-fitting lid. When filling storage containers, immediately sweep up any spills. Gather eggs from nests before nightfall.

Rats (but not mice) require a daily intake of water. Removing sources of drinking water during the night may be all that's needed to discourage rats from hanging around.

Turnabout is fair play; chickens prey on mice.

A raised coop floor should be at least 1 foot (30 cm) above soil level so rodents won't feel protected underneath. A coop floor consisting of dirt should be lined with ¼-inch (6 mm) hardware cloth, with attention to securing it tightly to corners and edges where the floor and walls meet. Better yet is a properly poured concrete floor. Secure vents and all other openings with ¼-inch hardware cloth.

Poisoning rodents is not the best idea. The poison itself is dangerous to any chickens, pets, wildlife, or children getting access to the bait. Some types of poisons pose a risk for a chicken or any other animal that might subsequently eat a poisoned rodent. And poisoned rodents tend to hide before they die, resulting in a horrific stench.

Traps are also problematic. Most traps require handling dead rodents, which is not a healthful plan for the person doing the handling. If a rat once sets off a baited trap but fails to get caught, it won't go near the trap again. Further, unless conditions that attract rodents aren't altered, new ones will take the place of those removed by trapping.

A cat that is not overfed may catch a fair number of mice, and maybe young rats, too, but likely won't have much interest in tangling with mature rats. A cat is less likely to prey on baby chicks if it has once experienced the ferocity of a protective mother hen.

A dog discipline-trained not to pester poultry will keep rats at bay. Especially popular for the purpose are terriers, dachshunds, and German pinschers.

Nearly all other predators relish rats and mice. On our farm we'll periodically see a rat snake in one of our barns, moving on once it has cleared out any resident rodents. One winter an eastern screech owl roosted under our barn roof and demonstrated, via regurgitated pellets and partially eaten carcasses beneath its perch, that it spent busy nights patrolling for rodents until it moved into spring digs.

Rats and Mice: A Closer Look

NORWAY RAT

APPEARANCE: stocky

WEIGHT: 10–18 oz (280–500 g)

MOVES: deliberately

CLIMBING: clumsy

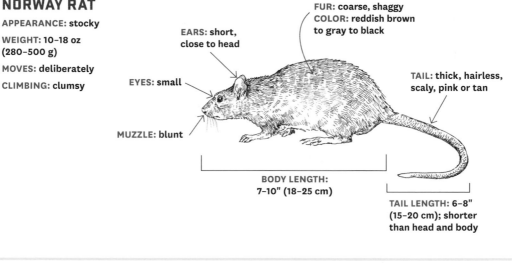

EARS: short, close to head

FUR: coarse, shaggy
COLOR: reddish brown to gray to black

EYES: small

TAIL: thick, hairless, scaly, pink or tan

MUZZLE: blunt

BODY LENGTH: 7–10" (18–25 cm)

TAIL LENGTH: 6–8" (15–20 cm); shorter than head and body

ROOF RAT

APPEARANCE: slender

WEIGHT: 6–12 oz (170–340 g)

MOVES: rapidly

CLIMBING: agile

EARS: large, creased, upright

FUR: sleek
COLOR: black to brownish gray

EYES: prominent

TAIL: thin with fine scales, dark gray, prehensile

MUZZLE: pointed

BODY LENGTH: 6–8" (15–20 cm)

TAIL LENGTH: 7–10" (18–25 cm); longer than head and body

HOUSE MOUSE

APPEARANCE: roundish

WEIGHT: ½–1 oz (14–28 g)

MOVES: rapidly

CLIMBING: good

EARS: large, round

FUR: smooth
COLOR: grayish brown

EYES: small

TAIL: thin, hairless

MUZZLE: pointed

TAIL LENGTH: 2½–4" (6–10 cm); same length as head and body

BODY LENGTH: 2½–4" (6–10 cm)

American Crow

AVERAGE LENGTH: 18 inches (46 cm)
AVERAGE WEIGHT: 1 pound (454 g)
AVERAGE WINGSPAN: 32 inches (81 cm)

As stated in their alternative name — common crows — American crows (*Corvus brachyrhynchos*) are among the most common birds in North America. Since they sleep at night and are active in the daytime, they are easy to spot. Their large size, jet-black plumage, and distinctive *caw* make them a cinch to identify.

Alert, observant, inquisitive, and smart, crows travel far and wide in search of vittles. A daily trip in summer might be 10 miles, while in winter the quest for eats might stretch to 30 miles. These birds are omnivorous and will consume just about anything that's remotely edible, including caterpillars, rodents, reptiles, amphibians, fruit, grain, garbage, and carrion. An average crow diet consists of two-thirds plants and one-third animals, the latter including the eggs and young of other bird species.

TYPICAL SIGNS. Crows are notorious egg snatchers. The eggs of outdoor-nesting ducks and guineas are easy picking, especially when vegetation is not dense enough to hide eggs. Where crows are habituated to people, they may become so bold as to march inside a coop to steal eggs right out of a nest box. Given the opportunity, they will also eat hatchlings.

A crow may prey on poultry solo, or with its mate, or with its extended family. The typical technique for stealing eggs is to snatch and run, enjoying the meal elsewhere. Small eggs may be carried whole between the two halves of the beak, leaving no evidence whatsoever.

When an egg is too big to carry intact, the crow pokes a hole in the shell, inserts its lower beak into the hole, and clamps down with its upper beak to carry the egg, keeping the hole tilted upward to avoid spilling out its contents. Evidence of crow predation is tiny pieces of shell in a nest where eggs are missing. Empty shells may be found as far as 100 yards from the nest, or sometimes even farther.

A large egg that's too heavy to carry may be eaten on-site, the empty shell left near the nest. The shell will most likely have a round or irregular hole in the side, with coarsely jagged edges.

HABITAT AND RANGE. American crows inhabit most of the United States and Canada, with these exceptions: far northern Canada, Alaska, Hawaii, the Pacific Northwest, and the southwestern United States. They are found in a wide variety of habitats but favor open landscapes (where they can feed on the ground) with scattered trees or small woodlots (for nighttime roosting, nesting, and general safety). They avoid grasslands with no trees and also the reverse: that is, large tracts of forest.

Outwitting Crows (Not!)

On our Tennessee farm, we keep guinea fowl together with chickens, hoping the guineas will learn to roost inside the coop at night and lay their eggs in the nest boxes we provide. At dusk, the guineas do follow the chickens into the safety of the coop, but some guinea hens insist on laying their eggs outdoors. Often the only way we know they have started laying each spring is by finding empty shells left by crows on our gravel driveway. We know it's the work of crows because occasionally we startle a crow enjoying breakfast in the driveway.

Then we begin our search to find the guinea nests, trying to get some of the eggs before the crows get them all. Those tricky crows have the advantage, however. They get up earlier in the morning and have time to perch in a nearby pine tree to watch for hens going to and from a nest. A guinea cock standing guard while his lady lays her morning egg is another tip-off that the crows surely learn to watch for.

As summer progresses and the vegetation gets thicker and taller, we have a better chance of beating the crows to the eggs. When we find a nest, we sometimes leave a few eggs behind. Otherwise the guineas will nest elsewhere and we'd have to start our search all over again. The crows, of course, watch us hunting for nests and gathering eggs. Learning where the nests are, and knowing we might leave some eggs, they later swoop down and wipe out any that remain. So, despite our best efforts, our hunt begins anew.

Our guineas do a good job of patrolling our pastures and woodlot for bugs, but as a result of their wide-range wandering they have a high rate of attrition, thanks to a variety of local predators. In a typical year we lose about half the guinea flock, so each spring we need to hatch enough keets to maintain a viable bug patrol. Some years it's a challenge to keep the wily crows from putting us out of the guinea-hatching business, but most years we manage to hatch more than enough keets for ourselves and to share with friends and neighbors.

Communal roosting occurs mainly during fall and winter, and the same roost may be used year after year. As time goes by, small flocks gradually congregate into one large flock. A roost may include thousands of local crows plus any visiting wintertime northern migrants. The largest flocks — sometimes numbering in the tens of thousands — occur in the prairie states, where grain is plentiful. Such large roosts put pressure on local resources, which can result in clever crows seeking breakfast in poultry yards.

Crows thrive around humans, where they can find plenty of garbage and other good things to eat, and they like urban areas because they are less likely to get shot or preyed on by great horned owls. While urban crows become habituated to humans, rural crows tend to remain more wary.

BEST DETERRENTS. Let's start with what doesn't work: fake owls, rubber snakes, and scarecrows. Oh, sure, they may work for a day or two, but crows are way too smart to be easily fooled. Frequently relocating these devices may lengthen the time they work, but eventually the crows will catch on.

Similarly, noisemakers, lights, and water sprays may work for a while — longer if used in combination and varied in location and intensity — but *only* for a while. Further, crows that get used to people and our noisy ways may not respond to scare devices at all.

So, what does work? First, make the place unattractive to crows. Since their primary purpose in being there is to look for snacks, remove readily available food, including poultry rations. If you feed your poultry outdoors, use a covered treadle feeder. Turkeys or geese in the poultry yard can be fierce about keeping crows at bay.

Covering the poultry yard with netting or hardware cloth is the most effective way to keep out crows. If the yard is too big to completely cover, crisscrossing wires over the top will discourage crows from entering. Things that flutter (cloth or mylar strips) or flash (DVDs or aluminum pie tins), hung from the overhead wires, serve as an additional deterrent.

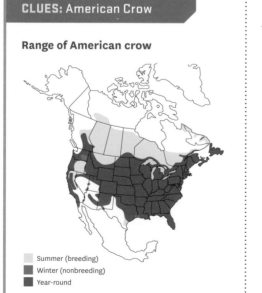

CLUES: American Crow

Range of American crow

- Summer (breeding)
- Winter (nonbreeding)
- Year-round

Pellet

A crow casts a pellet between four and eight hours after eating. Each pellet is about ½ inch (15 mm) long and ¼ inch (7 mm) in diameter and contains such things as undigested insect parts, fruit pits, tiny bones or bone fragments, and eggshells.

Tracks
FRONT AND REAR (L × W):
3¼ × 1⅜"
(8.4 × 3.5 cm)

Sound
Caw — strong and harsh

MOBBING

On our farm, we occasionally hear an out-burst of agitated cawing, rapidly accompanied by the sight of crows flying from all directions toward the commotion. Some unfortunate predator is being mobbed by a gang of crows. Most likely it's a great horned owl, having been spotted while trying to get some Zs. As the crows drive the sleepy owl from its perch, they call on their buddies to hasten forth and join the harassment, which may carry on for hours.

In this respect, crows can be beneficial to poultry keepers. Crows are preyed upon by many of the same animals that prey on poultry, and they attack a predator by cawing loudly while swooping at it. Public enemy number one is the great horned owl, which eats both young crows and adults. Enemy number two is the raccoon, which preys on eggs and hatchlings. Other species that crows may mob include barred owls, Cooper's hawks, sharp-shinned hawks, American kestrels, bald eagles, blue jays, ravens, vultures, red foxes, house cats, and even humans.

Mobbing serves several functions. It brings attention to the animal, ruining the predator's plan for a stealth approach. It distracts the predator from its intentions by keeping it busy defending itself. It helps young crows learn to identify predators and assess their degree of danger. Mobbing doesn't just draw attention to the mobbee — it also draws attention to the most macho crows at the head of the pack, the ones any sensible lady crows would consider as desirable mates.

But turnabout is fair play. A crow intent on raiding the nest of a blue jay or other songbird may itself be mercilessly mobbed.

Comparing Corvid Shape and Size

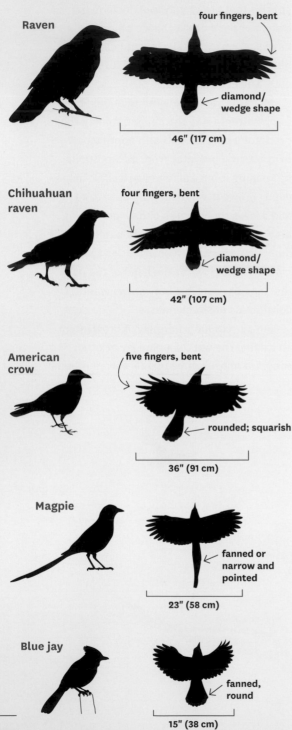

Raven

four fingers, bent

diamond/ wedge shape

46" (117 cm)

Chihuahuan raven

four fingers, bent

diamond/ wedge shape

42" (107 cm)

American crow

five fingers, bent

rounded; squarish

36" (91 cm)

Magpie

fanned or narrow and pointed

23" (58 cm)

Blue jay

fanned, round

15" (38 cm)

More Corvids

Crows and ravens are members of the genus *Corvus,* a Latin word meaning raven. They all look much alike and are easily confused with one another. They share these features in common:

- They are large perching birds (passerines).
- They are black from beak to feet.
- Their black feathers are glossy and iridescent.
- Bristle-like feathers cover their nostrils.
- They have stout legs.
- They have strong, heavy bills.
- They have large eyes and brains.
- They are intelligent and adaptable.
- They have awesome memories.
- They raise young cooperatively.
- They are gregarious.
- They have raucous voices.
- They are territorial.
- They cache eggs and other food.
- They'll eat anything remotely edible.
- They find most of what they eat while walking on the ground.

North America is home to three crow species and two raven species: American crow, fish crow, northwestern crow, common raven, and Chihuahuan raven. In most areas, no more than two species occupy the same range, somewhat simplifying identification. The most common combinations include the American crow, the range of which overlaps those of all four other *Corvus* species. Signs and deterrents are essentially the same for all five.

Common Raven

AVERAGE LENGTH: 25 inches (64 cm)
AVERAGE WEIGHT: 40 ounces (1.1 kg)
AVERAGE WINGSPAN: 46 inches (117 cm)

The largest of the perching species (passerines), the common raven (*Corvus corax*) is similar in size to a red-tailed hawk; even its appearance is somewhat raptorlike. The raven is larger and stockier than a crow and has a much stouter head, a long, heavier beak, stronger legs, and proportionally longer, narrower wings.

To identify a raven, look for its sometimes visible shaggy, beardlike throat feathers, and listen for its hoarse guttural croak. Also, unlike the crow's even-length tail feathers, the raven's tail feathers are longer in the middle, giving the tail a diamond shape when the bird is in flight. Compared to a crow's steady wing flapping and straight-path flight, the raven frequently soars or glides between occasional slow wing beats and enjoys launching into amazing aerial

acrobatics including rolls, somersaults, loops, and dives.

Ravens typically travel in pairs or in small family groups, but like crows they roost in large congregations in winter and may also gather together at landfills and other feeding sites. They are more carnivorous than crows and more typically may be seen gliding along highways with an eye out for roadkill. They also prey on rodents and the eggs and hatchlings of other bird species. Like a crow, a raven may leave a shell with a small round hole or carry away the entire egg. Having a larger, stronger beak than a crow, the raven can carry larger eggs.

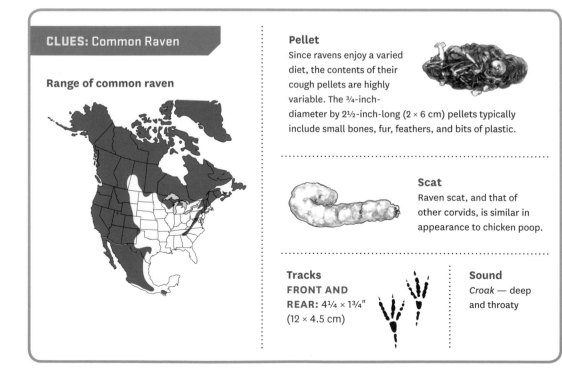

CLUES: Common Raven

Range of common raven

Pellet
Since ravens enjoy a varied diet, the contents of their cough pellets are highly variable. The ¾-inch-diameter by 2½-inch-long (2 × 6 cm) pellets typically include small bones, fur, feathers, and bits of plastic.

Scat
Raven scat, and that of other corvids, is similar in appearance to chicken poop.

Tracks
FRONT AND REAR: 4¼ × 1¾" (12 × 4.5 cm)

Sound
Croak — deep and throaty

Chihuahuan Raven

AVERAGE LENGTH: 19 inches (48 cm)
AVERAGE WEIGHT: 18 ounces (510 g)
AVERAGE WINGSPAN: 42 inches (107 cm)

Smaller than common ravens, Chihuahuan ravens (*Corvus cryptoleucus*) are slightly larger than American crows. Their caw is harsher than that of a crow, but less deep and throaty than that of a common raven. Like common ravens, Chihuahuan ravens have a diamond-shaped tail in flight and glide more than flap their wings. They can be distinguished from common ravens by voice and by their smaller size and their shorter, blunter beak with longer bristle-like feathers along the upper ridge.

Large insects are the favorite food of Chihuahuan ravens. They also like kitchen scraps, obtained by frequenting landfills, campsites, and schoolyards. Although they sometimes eat seeds, grains, and cactus fruit, they prefer carrion, lizards, small mammals, and the eggs and hatchlings of other birds, including poultry.

Magpies

AVERAGE LENGTH: 20 inches (51 cm)
AVERAGE WEIGHT: 6 ounces (170 g)
AVERAGE WINGSPAN: 23 inches (58.5 cm)

Also members of the corvid family, magpies (*Pica* spp.) are slightly smaller than crows, and more colorful. Their plumage is a stunning black and white, with iridescent blue-green wings and tail. A magpie's shape is similar to a crow's, but the tail has a diamond shape, like that of a raven, only proportionally much longer — often making up half the bird's total length. This feature enables the magpie to make sudden raptor-evading turns during flight.

Like crows and ravens, magpies forage during daylight and are omnivorous, although they have more voracious appetites for animal matter in the form of ground-dwelling arthropods, including caterpillars, grasshoppers, and beetles. Besides insects, they eat berries, seeds, nuts, food scraps, carrion (or fly maggots thereon), rodents, and the eggs and hatchlings of other birds.

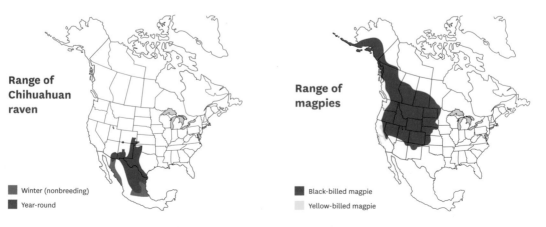

Range of Chihuahuan raven

■ Winter (nonbreeding)
■ Year-round

Range of magpies

■ Black-billed magpie
■ Yellow-billed magpie

Corvids: A Closer Look

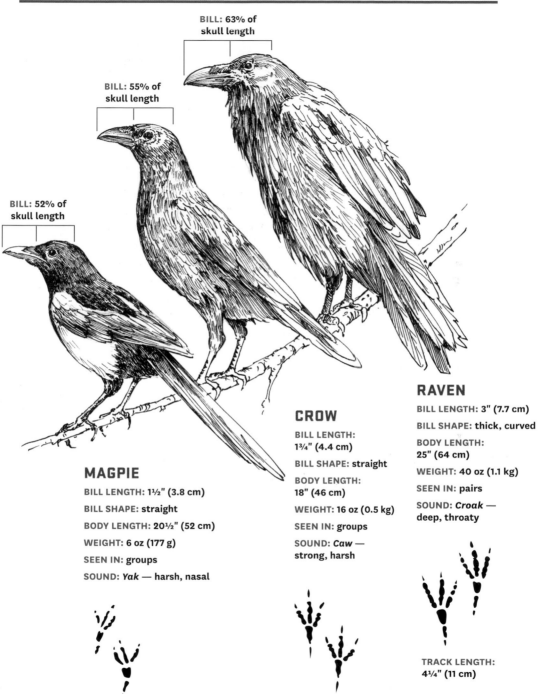

BILL: 63% of skull length

BILL: 55% of skull length

BILL: 52% of skull length

RAVEN

BILL LENGTH: 3" (7.7 cm)

BILL SHAPE: thick, curved

BODY LENGTH: 25" (64 cm)

WEIGHT: 40 oz (1.1 kg)

SEEN IN: pairs

SOUND: *Croak* — deep, throaty

TRACK LENGTH: 4¼" (11 cm)

CROW

BILL LENGTH: 1¾" (4.4 cm)

BILL SHAPE: straight

BODY LENGTH: 18" (46 cm)

WEIGHT: 16 oz (0.5 kg)

SEEN IN: groups

SOUND: *Caw* — strong, harsh

TRACK LENGTH: 3¼" (8 cm)

MAGPIE

BILL LENGTH: 1½" (3.8 cm)

BILL SHAPE: straight

BODY LENGTH: 20½" (52 cm)

WEIGHT: 6 oz (177 g)

SEEN IN: groups

SOUND: *Yak* — harsh, nasal

TRACK LENGTH: 2" (5 cm)

Jays

LENGTH: 10 to 12½ inches (25–32 cm)
WEIGHT: 2½ to 4¼ ounces (70–120 g)
WINGSPAN: 13¾ to 18 inches (35–45 cm)

Rounding out the North American members of the Corvidae family that have a culinary interest in poultry and their eggs are jays. Jays are slightly smaller than magpies and much more ubiquitous. Like other corvids, they are social, gregarious, clever, inquisitive, feisty, bold, and noisy. Also like fellow corvids — and chickens — jays have a complex social hierarchy and a vocal communication system. They easily adapt to being around humans and are readily drawn to backyard bird feeders.

Most, but not all, jays are blue, yet only one species is a blue jay. Those with blue plumage include the true blue jay of the eastern United States and Steller's and scrub jays of the West. Canada's gray jay is, well, gray. Some jays have distinctive crests on top of their heads, while others are crestless. Blue jays and Steller's jays have crests; scrub jays and gray jays do not.

Jays are mostly vegetarian, but like other corvids they eat just about anything. They are habitual nest raiders, targeting both eggs and hatchlings, including those of poultry. They may be deterred by the same methods used to deter magpies.

The best protection is a sturdy, completely enclosed coop and run. Otherwise, protectively house hatchlings until they are at least two weeks old. Collect eggs often. Hang fly curtain strips in pophole openings to discourage corvids from entering the coop to steal eggs.

Range of blue jays

■ Year-round
■ Winter (scarce)

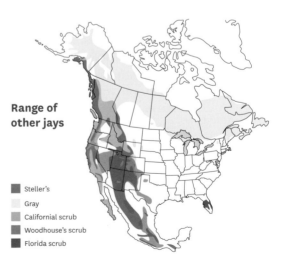

Range of other jays

■ Steller's
■ Gray
■ Californial scrub
■ Woodhouse's scrub
■ Florida scrub

Gulls

LENGTH: 13½ to 30 inches (34 –76 cm)
WEIGHT: 10 to 58 ounces (0.3–1.6 kg)
WINGSPAN: 3 to 5 feet (0.9–1.5 m)

Seabirds of the family Laridae, most gull species are about the size of a crow, and in fact gulls often compete with crows in scavenging for edibles in areas populated by humans. North America is home to more than a dozen gull species, which may be identified by their white and gray plumage with black markings, long, stout and slightly hooked beaks, long wings, webbed feet, and raucous squawking.

They are commonly called seagulls, although they don't all inhabit beaches and shorelines. Some live inland, preferring wetlands, ponds, rivers, and reservoirs, and may be spotted dining in farm fields, landfills, or parking lots at shopping malls and fast-food joints.

Gulls are gregarious, adaptable, and noisy. They are also clever hunters. If a gull startles a bevy of ducklings into diving, it simply waits for one to come up for air. A gull might crack the hard shell of a mollusk, or an egg, by dropping it onto a rock or other hard surface.

Like a snake, a gull can unhinge its jaw, allowing its beak to open wide enough for the gull to consume prey larger than its head. Like a snake, a gull may swallow a whole egg, shell and all. Or, like other birds, it might peck a hole in the shell. Or it might use its hooked beak to bite into the shell. The gull will typically eat the contents, at least partially, before carrying the shell elsewhere. If you find the shell, it will likely contain yolk residue.

All gull species are opportunistic feeders known to raid outdoor nests for eggs and hatchlings. They may even attack mature poultry, especially any that seem weak, ill, or injured. Avoiding predation by gulls is basically the same as for corvids.

◄ Gulls can be difficult to identify by species, especially where their ranges overlap and they interbreed. Further, gulls don't achieve adult plumage until the age of 3 or 4, and until then their plumage patterns change each year. Even as adults, some species have one color pattern while breeding and another during non-breeding season. Once you've seen a gull, however, you won't have trouble recognizing one next time.

REPTILES AND AMPHIBIANS

Rat Snakes · More Colubrids · American Bullfrog · Snapping Turtles ·
Alligators and Crocodiles

Rat Snakes

Gray rat snake

LENGTH: 5 to 8 feet (1.5–2.4 m); varies by species

Found throughout the United States and into southern Ontario, rat snakes go by many names, one of which is chicken snake, and come in several types. They are the most widespread of all North American snakes that dine on poultry eggs and hatchlings.

All rat snakes are nonvenomous. They are medium to long constrictors, meaning that before they eat live prey they kill it by squeezing until the circulatory system ceases to function and blood can no longer reach the victim's brain. As implied by their name, rats are their favorite delicacy.

Aside from relishing rats and other rodents, as well as birds and bird eggs, rat snakes eat frogs, lizards, and other snakes, including rattlesnakes and copperheads. The ability to control venomous snakes is one of many reasons I'm always happy to see a rat snake on patrol.

If you are not familiar with the snake species in your area, start by learning to identify the ones that are venomous. In most areas, venomous snakes are far fewer than nonvenomous ones. On our farm, for example, we watch out only for copperheads and rattlers. Both are much stouter than our slender, nonvenomous snakes and both have distinctive identifiable markings. The more rat snakes we see, the fewer copperheads and rattlers we encounter.

TYPICAL SIGNS. Rats and other rodents tend to hang around coops and barns, making powerful attractants for a rat snake. When the rodents are cleared out, the rat snake will look around for something else interesting to eat. Bird eggs are a common food for rat snakes, and poultry eggs are no exception. Young poultry, up to about one month of age, are also attractive because they are small enough to swallow and don't put up much of a fight, especially at night.

While a snake does have teeth, they are designed for grabbing and holding prey, not for chewing, so a snake swallows its prey whole. It can unhinge its jaw to eat things bigger than its head, but sometimes a snake gets overly ambitious. A dead half-grown chick with damp and matted feathers on its head and neck is likely a victim of a snake with eyes bigger than its stomach. I once found a damp and dead young

Silkie with one wing sticking out at a right angle, apparently having blocked the bird's passage down a snake's throat. Unable to get the chick down, the snake spit it back out.

Identifying predation by a rat snake otherwise can be difficult because the snake eats a chick or egg whole. The only sign is therefore one or more chicks or eggs missing. A snake that has slithered into a coop through a tiny opening and then fattened up on chicks or eggs may not be able to slip back out the way it came in. And a snake that becomes sated after eating multiple chicks or eggs may be loath to move. In either case, you may find the snake coiled up in a nest or a corner of the coop, sleeping off its latest meal.

When the snake feels like traveling, it'll typically move along on its own. It won't eat again until it has finished digesting an egg, chick, or rodent, which takes about four days. Meanwhile, a snake that isn't hungry has little incentive to stick around an active poultry yard. Most rat snakes will be gone within a day.

APPEARANCE, HABITAT, AND RANGE. Common lengths for an adult rat snake range between 4 and 6 feet, although individuals can grow as long as 10 feet. Climate affects length; snakes in warmer areas tend to be shorter.

Rat snakes may be seen on the ground, swimming in water, or climbing or basking in a tree. They may also hang out in the rafters of a coop or barn and will just as readily enter a house. A video on YouTube shows a black rat snake entering a North Carolina home to pilfer chicken eggs from the kitchen counter.

A rat snake has a small home range. We repeatedly see the same snakes, or their shed skins, in the same spots in our backyard, garden, or barn. The adult appearance of each population varies throughout its geographic range, and populations interbreed.

Yellow rat snake

Great Plains rat snake

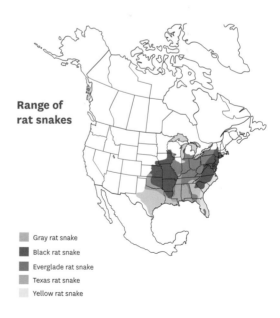
Range of rat snakes

Gray rat snake
Black rat snake
Everglade rat snake
Texas rat snake
Yellow rat snake

Life with Rat Snakes

As part of the morning routine I let the Khaki Campbell ducks out of their nighttime coop and then lifted the exterior lid to the row of nest boxes to gather their eggs. I was only half paying attention when I reached into the second nest and was startled by a big black rat snake, bulging with duck eggs.

It was my first encounter with rat snakes, but far from the last. Over the years, we have seen plenty of rat snakes in nest boxes, slithering around the floor of our barn, high atop stacked hay bales, or prowling the garden. Most of the time, I ignore them and they ignore me.

I did not ignore the rat snake that once managed to get inside a coop full of growing chicks. Before we repaired the gap it had squeezed through to get in, we opened the door and encouraged the snake to leave. As soon as it was out in the open, our flock of guinea fowl caught sight of it and harassed it until it had found cover away from the building. I don't believe that snake, after being terrorized by the outraged guineas, was ready to return any time soon.

Another time I found a 6-footer in our brooder area. Unable to gain access to the chicks through the quarter-inch hardware cloth, it slithered across the floor, up the wall, and into the goats' hay manger, where one of our New Hampshire hens had laid an egg. I watched, spellbound, as it slowly worked the large egg into its mouth and down its throat. I was so engrossed that when the snake squeezed the egg and broke the shell with a loud *snap*! I jumped, as if I'd been shot.

Whenever we see a rat snake in our barn, we know it's doing a good job of clearing out pesky rodents. Occasionally one will pilfer an egg or two, but that's okay, because we typically have more than we need. And it's a small price to pay for rarely encountering rats, mice, copperheads, or rattlesnakes.

BEST DETERRENTS. Excluding rat snakes poses a considerable challenge. They can squeeze through small openings and climb walls to find them. Covering all openings with ¼-inch (6 mm) hardware cloth will exclude snakes at night, but not during the day when they could slither in through the open pophole. During the daytime, active poultry — guineas, turkeys, geese, and ducks (especially Muscovies) — will discourage snakes from hanging around, both because snakes avoid hustle and bustle and because poultry sometimes attack and even eat snakes.

Remove hiding places around the coop, such as piles of debris, wood, or rocks. Keep grass mowed. Collect eggs often, and especially don't leave them in nests overnight. Check the coop for snakes before locking up at night. Make an effort to control rodents. Even though rat snakes will help you out there, they also will use rodent tunnels to get inside the coop, so fill tunnels as you find them.

If you are desperate to keep out snakes, you might erect a snake fence. It will be difficult to maintain (keeping it clear of leaves and other blown debris), but it'll do the job. This fence is constructed of ¼-inch hardware cloth at least 30 inches high and buried 6 inches into the ground. The posts are angled 30 degrees outward, so snakes can't get a grip on the fence, and set on the poultry yard side of the hardware cloth, so snakes can't climb on the posts. This fence must be checked often to make sure no branches or other debris have blown against it that a snake could use to climb on. Since the fence won't keep out most other predators, or keep in most poultry, it is added to the outside of the regular poultry fence and gates.

Snake repellents work to varying degrees, from temporarily to not at all. Some require frequent application, may be toxic, or are so smelly that they are more likely to repel humans than snakes. Traps work better, although you'll need to check them often and release any captured snake to a location 200 yards or less from where it was trapped. Although the snake will still be within its home range, it will associate the bad experience of being moved with the place where it was trapped and is not likely to return to that spot. Above all, remember that rat snakes are not only harmless but beneficial, and some species are endangered, largely thanks to unenlightened humans.

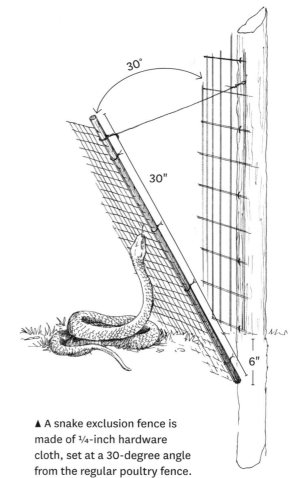

▲ A snake exclusion fence is made of ¼-inch hardware cloth, set at a 30-degree angle from the regular poultry fence.

Copperhead

Corn snake

RAT SNAKE OR COPPERHEAD?

All young rat snakes have vivid light and dark blotches in a pattern that is easily mistaken for that of a young copperhead. A baby copperhead, however, may be readily distinguished by its bright yellow tail tip, a feature it wiggles in imitation of a worm or caterpillar to entice frogs, lizards, and other delicacies to get close enough to become lunch. As the copperhead reaches one year of age, the yellow tail gradually fades into brown. Meanwhile, the blotches of most rat snakes also fade as they mature into their adult pattern.

Corn snakes and fox snakes, however, maintain their blotchy appearance and are often confused with venomous snakes. Fox snakes are sometimes mistaken for copperheads, which is odd because copperheads do not share the same range. However, the fox snake shares range with the look-alike massasauga rattlesnake and will similarly hiss and shake its tail when threatened, but it lacks rattles.

Corn snakes and copperheads share much the same range, and both have keeled scales, so that's not a helpful trait for identification. A corn snake, however, is lighter in color, with squarish blotches that do not extend all the way down the sides, and a narrow wedge-shaped head with a V pattern on top. A copperhead has a stouter body, hourglass-shaped dark blotches that extend down the sides, and an arrow-shaped coppery-brown head. Several other distinctions (differences in pupil shape, the presence or absence of heat-sensing pits, and the arrangement of scales under the tail) may be determined only by close inspection that could put you in danger if indeed you are examining a copperhead. So concentrate on looking at the snake's body thickness and blotch shape from a distance.

Some rat snakes, especially red rat snakes (a.k.a. corn snakes), are docile and easy to handle. Others, especially Texas rat snakes and some black rat snakes, are decidedly aggressive and will bite if they feel threatened. Although they do not have venom-filled fangs, the bite can be painful. Further, bacteria in the saliva can cause a serious infection.

In most cases, a rat snake will strike and then release. If it hangs on, the first order of business is to get its curved teeth out of your skin. A natural reaction is to try to pull the snake away — a bad move that will tear your flesh and possibly break the snake's teeth. Instead, take a firm hold behind the snake's head and push the head toward the bite until its teeth come out of your skin.

Assuming you are confident that you were not bitten by a venomous snake (in which case, hasten to the ER!), treat the bite as you would any superficial skin wound. Start by thoroughly rinsing the wound with plain water, preferably running from a tap. Next, wash the skin around the wound with soap, avoiding getting soap into the wound.

When the wound and surrounding area have been thoroughly cleaned, blot dry. Apply Green Goo salve (my favorite), petroleum jelly, or an antibacterial wound ointment such as Neosporin. Leave the wound open to the air, unless it needs to be temporarily bandaged to remain sanitary (for instance, while you clean out the coop). Watch for signs of infection — redness, swelling, warmth, leaking, increased pain — requiring a visit with your family doctor.

More Rat Snakes and Their Ranges

Range of corn snakes

■ Red corn snake
■ Great plains snake

Range of fox snakes

■ Eastern fox snake
■ Western fox snake

More Colubrids

Colubrids are members of the Colubridae family of snakes, a large and diverse group that represents some two-thirds of all known snake species. After rat snakes, members most likely to relish poultry eggs or chicks are gopher snakes, bull snakes, and pine snakes (collectively members of the genus *Pituophis*), king snakes, and racers — all members of the large subfamily Colubrinae.

Like rat snakes, they are all nonvenomous, but they can inflict a painful bite. When annoyed they shake their tails, like a rattlesnake. If threatened, they may emit an unpleasant-smelling musk. Their primary diet consists of rodents, so of course they are attracted to rodent-infested chicken coops.

Scarlet king snake

Southern black racer

Range of *Pituophis* snakes

- Pine snake
- Gopher snake
- Bull/gopher hybrids
- Bull snake

Range of racers

- Northern black racer
- Black-masked racer
- Blue racer
- Buttermilk racer
- Eastern yellow-bellied racer
- Tan racer
- Western yellow-bellied racer
- Southern black racer

Range of king snakes

- Black king snake
- California king snake
- Eastern king snake
- Desert king snake
- Speckled king snake

American Bullfrog

AVERAGE LENGTH: male, 6 inches (15 cm); female, 6½ inches (17 cm)
AVERAGE WEIGHT: 1 to 1½ pounds (0.5–0.7 kg)

Our largest native North American frog, the American bullfrog (*Lithobates catesbeianus*) has strong, powerful legs, a large head, and a wide mouth into which it stuffs all manner of things including insects, worms, mice and other small land mammals, birds, bats, snakes, turtles, fish, and other frogs. It is solitary, wary, and primarily nocturnal.

Mature bullfrogs eat only live food. They are carnivorous and cannibalistic — bullfrogs sometimes comprise as much as 80 percent of a bullfrog's diet.

TYPICAL SIGNS. The most obvious sign that bullfrogs are around is hearing the low-pitched resonant bellow of males — *Rumm, R-R Rumm* — that has been compared to the roaring of a bull (hence the name bullfrog). Throughout warm spring and summer months it continues day and night and carries for up to half a mile.

Signs of bullfrog predation include the disappearance of ducklings or goslings, or maybe the discovery of a bullfrog that choked to death trying to swallow a slightly-too-big duckling or gosling. A bullfrog sits and waits for an opportune moment and then uses its muscular hind legs to lunge, lightning fast, toward prey, mouth open wide and sticky, elastic tongue unfurling to snag animals up to twice the frog's own weight. Any live animal a bullfrog can cram into its mouth is a potential meal.

APPEARANCE. The tailed tadpoles are long and thin, dark green with black spots, a yellowish underside, and orange or bronze eyes. Gradually they grow legs and then arms, and the tail starts to absorb into the body. The tadpole's gills recede as the frog develops lungs that allow it to venture out of water onto land.

The back of a fully developed bullfrog is rough, peppered with tiny bumps, and typically green to gray brown with dark spots. Color variations include dark green with light green spots, light green with yellow spots, and brown with white spots. Some have a distinct netlike pattern. The underside is grayish. The male has a bright yellow throat, while the slightly larger female has a white throat. Males and females both have blunt-tipped fingers and toes with well-developed webbing. The male, however, has a larger thumb and forearm than the female.

Both have golden eyes and a tympanum, or circular eardrum, on each side of the head, just behind the eye. In females, the tympanum is no bigger than the eye; in males, it is twice the size of the eye. A ridge, or fold of skin, runs from the eye around the tympanum toward the mouth. The similar equally voracious, but slightly smaller green frog (*Lithobates clamitans*) may be distinguished from the bullfrog by these two ridges, which in the green frog continue down the length of the back.

The average length of male bullfrogs is about 6 inches and females are about 6½ inches. Average weight is a tad more than 1 pound, topping out at 1½ pounds.

HABITAT AND RANGE. Bullfrogs are native to the eastern half of North America, ranging from Nova Scotia to central Florida and the Gulf of Mexico, and westward into eastern Wyoming, Colorado, and New Mexico. They have been widely introduced into western areas as a biological control for mosquitoes and other pests, as released or escaped pets, and for culinary purposes. Many people consider frog legs to be a delicacy. They are about on a par with chicken wings, with a slightly fishy flavor.

Where bullfrogs have been introduced they have become a seriously problematic invasive species thanks to their broad climatic tolerance, their prolific reproduction, and their voracious appetite for native frogs and other species. Further, they outcompete native species for space and food by aggressively defending their newly established territories against them. The Global Invasive Species Database includes the American bullfrog on its list of "One Hundred of the World's Worst Invasive Alien Species."

BEST DETERRENTS. Wherever bullfrogs are found, their populations are increasing. Nowhere are conservation efforts made on behalf of this species. Rather, where bullfrogs are not native, eradication is a conservation priority. "We get a lot of complaints from landowners whose ducks are getting picked off by bullfrogs," says Leanne Leith, wildlife coordinator with the Langley Environmental Partners Society in British Columbia, where bullfrogs have been introduced

The incredible mobility that allows the species to spread far and fast also makes bullfrogs difficult to catch by hand. Further, they can be aggressive and bite, although without doing serious damage. Eradication options include shooting with a firearm, airgun, or bow and arrow, spearing, or hooking with a fishing pole. In some states, a license is required.

Personally, I like having bullfrogs in our farm ponds. Despite the sometimes annoying and constant bellowing of the males, they do such a good job of keeping down mosquitoes that we can enjoy quality time on our backyard deck. And it's not so difficult to keep fast-growing ducklings and goslings confined away from the pond until they are too big for a bullfrog to cram into its massive maw.

Range of American bullfrog

STATE ANIMALS

The American bullfrog is the state amphibian of both Oklahoma and Missouri and the official frog of Ohio. The northern black racer is Ohio's state reptile. The common snapping turtle is New York's state reptile. The American alligator is the state reptile in Florida, Louisiana, and Mississippi.

Frog Tale — No Bull

Frieda McArdell, Illinois, owner-operator, Passion for Ponds (www.passionforponds.com)

Bullfrogs are a good example of something that is exciting to find in your pond for the first time. I have had one or two bullfrogs in my pond since I put the water in. They just showed up. There isn't any large water source within 5 miles, but somehow they found their way here. I never gave much thought to them being here. After all, they eat bugs, and that's a good thing, right?

On Memorial Day, my husband's entire family descends on our house and we have a big party, so I thought it would be a perfect opportunity to introduce the baby mallard to the pond. I had rescued the duckling three days earlier as the sole survivor of a predator attack in a client's courtyard. Plenty of people would be around to make sure the duck didn't run amuck (he-he!), get under the deck, or, God forbid, have one of the cats get it. So in the pond he went.

He was the cutest little thing! He sat up on the edge of one of my plant pots and preened, then he would jump back in the pond and swim around like he was having the time of his life! Everyone there couldn't get over how cute he was. Then, unexpectedly, it happened.

One of the young cousins shrieked, "A frog just ate the baby duck!"

My oldest son, Ryan, being a quick thinker, grabbed the net at the pond edge and quickly netted the frog out of the pond with the duck in tow. In the confusion, the frog escaped. By then I was pondside and there was no way that frog was getting away from me!

I trapped the frog and pried the lifeless baby duck from its mouth, hoping to revive it, but to no avail. Needless to say, what started as a happy day ended in tragedy.

The moral of the story? Bullfrogs will eat, or try to eat, anything they think they can fit in their mouths.

Snapping Turtles

LENGTH OF SHELL: 10 to 26 inches (25–66 cm); varies by species

WEIGHT: 23 to 200 pounds (10.4–91 kg)

Snapping turtles are large turtles that inhabit the eastern and southeastern United States and southeastern Canada. They are distinct from cute little box turtles in their intimidatingly huge size, in being aquatic rather than terrestrial, in lacking the ability to retreat entirely inside their shell, and in having a vicious bite.

Snapping turtles are members of the family Chelydridae. In North America we have two genera: one species of common snapping turtle (*Chelydra*) and three species of alligator snapping turtles (*Macrochelys*).

The common name *snapping turtle*, or *snapper*, describes this animal's powerful snapping jaws, which are used both for self-defense and for obtaining food. They can bite hard enough to remove one of your fingers (or whole hand), but they rarely attack unless teased or otherwise threatened.

Snapping turtles are sit-and-wait hunters. Their favorite meals are fish and aquatic plants — as my husband and I discovered when we stocked our farm pond with fingerling catfish and water lilies that rapidly vanished. Snapping turtles also eat carrion, frogs, snakes, smaller turtles, small mammals, and waterfowl.

BEST DETERRENTS. Snapping turtles won't climb over a fence or dig underneath, making a well-built poultry fence your best defense. They don't actively hunt poultry on land but may snag any poultry foraging near

Common snapping turtle

Alligator snapping turtle

Range of common snapping turtle

Range of alligator snapping turtle

water or visiting water's edge for a drink, or waterfowl swimming in the water.

When ducks or geese become nervous about entering their pond or stream, a snapper has likely taken up residence there. Typically the first sign of snapping turtle predation is the disappearance of ducklings or goslings. Another sign is seeing a parent duck or goose fighting off "something" in the water to defend offspring.

Snappers typically attack a swimming duck or goose by biting onto a leg and dragging the bird underwater before eating it. A grown duck or goose would be lucky to get away with only bruises. Waterfowl in snapper country are at risk of losing a leg, suffering a bill injury, or losing their lives.

A snapper on land can't move fast and typically won't stay long, so the best plan of action is to steer clear. In most states, you can relocate a snapping turtle that's causing poultry predation, although to avoid injury you'd be wise to get expert help from a wildlife removal service.

Snapping Turtles: A Closer Look

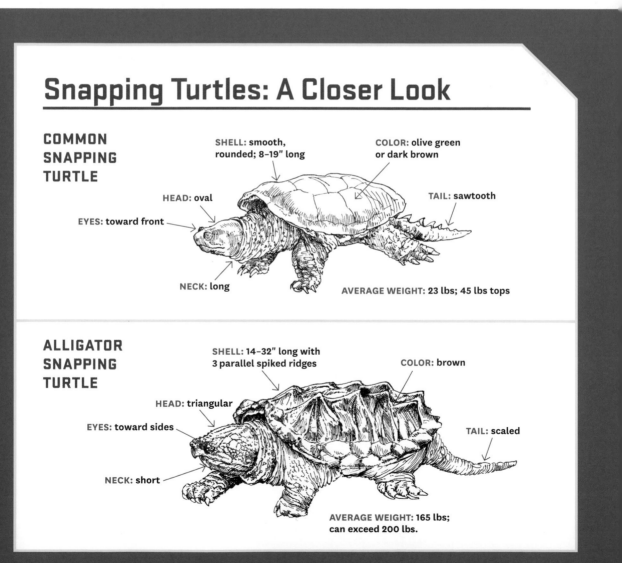

COMMON SNAPPING TURTLE

SHELL: smooth, rounded; 8–19" long

COLOR: olive green or dark brown

HEAD: oval

EYES: toward front

TAIL: sawtooth

NECK: long

AVERAGE WEIGHT: 23 lbs; 45 lbs tops

ALLIGATOR SNAPPING TURTLE

SHELL: 14–32" long with 3 parallel spiked ridges

COLOR: brown

HEAD: triangular

EYES: toward sides

TAIL: scaled

NECK: short

AVERAGE WEIGHT: 165 lbs; can exceed 200 lbs.

Alligators and Crocodiles

LENGTH (including tail): 8 to 13 feet (2.5–4 m); varies by species and gender
WEIGHT: 200 to 840 pounds (91–381 kg)

Crocodilians are members of the order Crocodylia, Earth's largest living reptiles. Two species are native to the United States — American alligators and American crocodiles.

Both species will eat whatever they can catch and swallow, including insects, crustaceans, fish, mammals, birds, amphibians, and reptiles.

Both alligators and crocodiles will eat ducks and geese swimming in water or chickens free-ranging near water. They prefer to eat things they can swallow with one or two gulps. A young alligator or crocodile may snack on ducklings and goslings, but an older one may also take an interest in grown poultry.

Gators actively hunt in warm weather, mainly in the water and at night. They can swim as fast as 20 miles an hour, moving much faster in water than on land. In dry conditions,

FAMILY TRAITS

Despite their many differences, American alligators and crocodiles are similar enough to be confused with one another. Similarities include these:

- They look like giant lizards with long, muscular bodies.
- They have tough, scaly hides armored with bony plates.
- Their eyes and nostrils are on top of their head.
- Because they are cold-blooded, they prefer warm climates.
- They tend to be more active at night.
- They dig burrows for protection, shelter, and hibernation.
- They hibernate in cold weather.
- They sunbathe to get warm and swim to cool off.
- They produce young by laying eggs.
- Incubation temperature determines the sex of offspring.
- They live in or near water and are excellent swimmers.
- The flat, muscular tail is about half the length of the body.
- They have four short legs for walking on land.
- They have five toes on the front feet and four on the back feet.
- They have excellent vision and keen hearing.
- They have powerful jaws and formidable teeth.
- They use their teeth for biting and tearing, rather than chewing.
- They sometimes stash prey underwater to eat later.
- They are carnivores with similar diets.
- They swallow stones to help grind food in their stomachs.
- They have been on Earth, unchanged, for millions of years.

American alligator

American crocodile

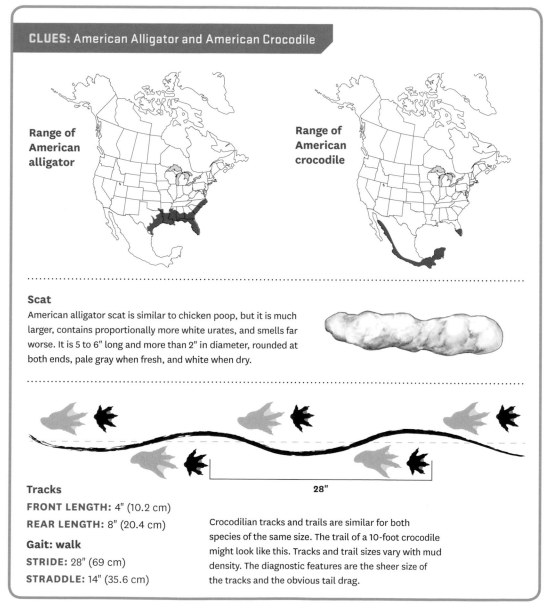

CLUES: American Alligator and American Crocodile

Range of American alligator

Range of American crocodile

Scat

American alligator scat is similar to chicken poop, but it is much larger, contains proportionally more white urates, and smells far worse. It is 5 to 6" long and more than 2" in diameter, rounded at both ends, pale gray when fresh, and white when dry.

28"

Tracks
FRONT LENGTH: 4" (10.2 cm)
REAR LENGTH: 8" (20.4 cm)

Gait: walk
STRIDE: 28" (69 cm)
STRADDLE: 14" (35.6 cm)

Crocodilian tracks and trails are similar for both species of the same size. The trail of a 10-foot crocodile might look like this. Tracks and trail sizes vary with mud density. The diagnostic features are the sheer size of the tracks and the obvious tail drag.

gators cross land and even enter suburbs looking for suitable water to inhabit. During the day, they tend to be sit-and-wait hunters, loitering on land to snag creatures that wander to the water's edge for a drink.

BEST DETERRENTS. The best way to protect poultry from gators and crocs is to house the flock within a well-built fence. Although gators may climb a fence when threatened, they do so out of desperation rather than to seek an avian meal. A single hot wire a few inches above the ground (snout height to a crocodilian) will provide added protection. Although repellents touted online may or may not be effective, they are likely to offend you and your neighbors.

Water attracts gators. If you have a swimming pool or water garden, fence it in. Potential food sources in the yard attract gators. Put away the barbecue grill when it's not in use. Fence not just poultry but also pet rabbits, rodents, snakes, and fish. If any such animals are on your property as wildlife, don't encourage them to stay by feeding them. And for heaven's sake, don't feed gators, which gives them little incentive to leave.

If a gator hangs around your place, get professional help. Call your local animal or wildlife control agency, or call 911 — the dispatcher will know where to get help.

BE SAFE!

Crocodilians eat by capturing and crushing prey with their immense jaws and then swallowing the meal whole. Their strong stomach acid digests soft tissue in two to three days. Bones and other hard parts take longer — sometimes as much as eight months.

An American crocodilian's jaw can apply 2,000 to 3,000 pounds of pressure per square inch (psi), depending on the animal's size. The bigger the crocodilian, the stronger its bite force. Compare that to a human's biting strength, which tops out at 200 psi.

Chomping down on the likes of a tough snapping turtle shell can result in broken teeth. Not a problem — a crocodilian's teeth are replaceable. Each gator or croc goes through thousands of teeth in its lifetime.

A gator typically won't attack unless it's hungry or provoked, and crocs are shy and generally avoid humans. Most attacks occur when a human enters a crocodilian's watery domain or walks or squats along the water's edge at night. If you should chance upon an alligator or crocodile on land, especially during nesting season, the best way to avoid losing a hand or an arm is to steer clear.

An aroused crocodilian can accelerate fast. Although it can't sustain speed, it can outrun a human for about the first 20 feet, according to Jack Hanna, director emeritus of the Columbus (Ohio) Zoo and Aquarium. He recommends keeping a distance of at least 50 feet. Attacks — human or poultry — most often occur in and around water, so the safest plan is to keep yourself and your flock distant from potential crocodilian hazards, especially at night.

Alligators and Crocodiles: A Closer Look

AMERICAN ALLIGATOR

PREFERRED HABITAT:
fresh water

COLOR: nearly black

SNOUT: wide, rounded, U-shape

AVERAGE MATURE WEIGHT:
male, 500 lb (227 kg);
female, 200 lb (91 kg)

TEETH: lowers not visible with mouth closed

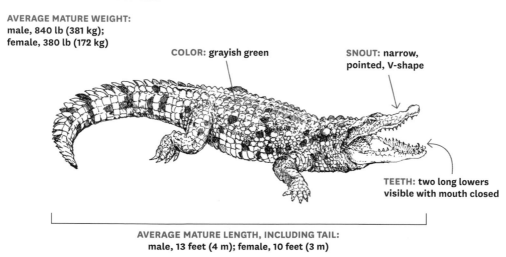

AVERAGE MATURE LENGTH, INCLUDING TAIL:
male, 11 feet (3.4 m); female, 8 feet (2.5 m)

AMERICAN CROCODILE

PREFERRED HABITAT: saline water

AVERAGE MATURE WEIGHT:
male, 840 lb (381 kg);
female, 380 lb (172 kg)

COLOR: grayish green

SNOUT: narrow, pointed, V-shape

TEETH: two long lowers visible with mouth closed

AVERAGE MATURE LENGTH, INCLUDING TAIL:
male, 13 feet (4 m); female, 10 feet (3 m)

Checklist: Predators in My Area

In the event of an attack on your poultry, this table will help you move quickly to identify the culprit. Using information and maps from the predator profiles earlier in this book, plus data from the many citizen science websites that track verified sightings of specific species (found through a keyword search for "citizen science" and the name of the species you are investigating) and any information available to you about local wildlife:

- Place a check mark in the "probable" column for each predator known to be in your area.

- Place a check mark in the "possible" column for each predator that may be in your area (for instance, if your area is not far from the animal's usual range).

- In the "season" column, indicate which months a migratory predator is likely to be in or pass through your area.

- Do not mark any column for predators unlikely to be in your area (based on known range).

PREDATOR	PROBABLE	POSSIBLE	SEASON
Alligator			
Badger			
Bear			
Bobcat			
Bullfrog			
Cat			
Coati			
Cougar			
Coyote			
Crocodile			
Crow			
Dog			
Eagle			
Falcon			
Fisher			
Fox			
Ground squirrel			
Gull			
Hawk			
Jay			
Lynx			
Magpie			
Marten			

PREDATOR	PROBABLE	POSSIBLE	SEASON
Mink			
Opossum			
Otter			
Owl			
Raccoon			
Rat			
Raven			
Ringtail			
Skunk			
Snake			
Snapping turtle			
Weasel			
Wolf			
Wolverine			

Identifying Predators by Bite Marks

Punctures left by canine teeth can provide helpful clues to a predator's identity. To assess puncture marks, examine both the underside of the skin and the flesh beneath the skin of a dead bird. The larger, deeper, and farther apart paired punctures are, the larger the predator.

Average Distance between Canine Teeth Punctures (Greatest to Least)

PREDATOR	IMPERIAL	METRIC
Bear, brown	2–3"	51–76 mm
Bear, black	1½–2½" 1³/₁₆–2 ⁷/₁₆	45–65 mm
Cougar	1½–2¼"	38–57 mm
Wolf	1¼–2³/₁₆"	32–55 mm
Coyote	1⅛–1⅜"	29–35 mm
Otter	1¹/₁₆–1¼	27–32 mm
Badger	⅞–1⅛"	22–28 mm
Bobcat	¾–1"	19–25 mm
Fox, red	¹¹/₁₆–1"	18–25 mm
Fox, gray	½–¾"	13–19 mm
Mink	⅜–½"	10–13 mm
Weasel, long-tailed	⁵/₁₆–⅜"	8–9 mm
Weasel, short-tailed	⅛–³/₁₆"	3–5 mm

Poultry Preferences of Predators

This table indicates typical preferences of each poultry predator species. Note, however, that individual predators may act in unpredictable ways. As soon as someone insists a certain predator "does not do that," someone else is certain to post a video on YouTube proving the opposite.

Predatory Preferences

PREDATOR	LARGE BIRDS*	MATURE BIRDS†	YOUNG BIRDS	EGGS	RANGE	COOP
Alligator	X	X	X		FW	
Badger	X	X	X	X		X
Bear		X		X		X
Bobcat	X	X	P	P		
Bullfrog			X		FW	
Cat			X	P		X
Coati		X	X	X		X
Cougar	X	X				
Coyote	X	X	X	X		
Crocodile	X	X	X		SW	
Crow			X	X		X
Dog	X	X	X	X		X
Eagle	X	X				
Falcon	X	X	X			
Fisher	X	X	X	X		X
Fox	X	X	X	X		X
Ground squirrel			X	X		
Gull			X	X		
Hawk		X	X			P
Jay			X	X		X
Lynx	X	X	P	P		
Magpie			X	X		X
Marten		X	X	X		X
Mink	X	X	X	X		X
Opossum		X	X			
Otter		P	X	X		X
Owl	X	X	X	X		
Raccoon	X	X	X	X		X
Rat			X	X		X
Raven		P	X	X		X
Ringtail		X	X	X		X
Skunk		P	X	X		X
Snake			X	X		X
Snapping turtle			X		FW	
Weasel	X	X	X			X
Wolf	X	P				
Wolverine	X	X	P	X		

*Large birds = turkeys, geese, and the largest duck and chicken breeds

†Mature birds = guineas and average and smaller breeds of ducks and chickens

X = Likely
P = Possible

FW = in or near fresh water
SW = in or near saline or brackish water

Glossary

accipiter. A genus of raptors in the family Accipitridae; generally called hawks

Accipitriformes. Members of a taxonomic order that includes the diurnal birds of prey except falcons

ball walker. An animal that walks on its toes (digitigrade); also called toe walker

binocular vision. Both eyes see the same object at the same time

bird of prey. A raptor

bound. An elongated gait similar to a hop, but with the hind foot prints behind the front prints

buteo. A hawk of the *Buteo* species (called "buzzard" in Europe)

buzzard. Common name for a turkey vulture

cache. A carcass or eggs partially or entirely hidden under soil or debris

carrion. A decaying flesh of dead or dying animals

casting. The process of coughing up a dry pellet of matter undigestible by a wild bird; also, the coughed-up pellet; also, the process of making a plaster mold of an animal's track

cathemeral. Active 24/7

chicken hawk. Common name for a Cooper's hawk, a sharp-shinned hawk, or a red-tailed hawk

chyme. Pulpy, acidic digestive material passed to the small intestine by a wild bird's gizzard

civet cat. Alternative name for either a ringtail or a spotted skunk

committee. A group of vultures roosting

corvid. A bird in the crow family

crepuscular. Most active during the twilight hours of both dawn and dusk [also called bimodal or bimodally active]

dead short. A direct connection between a fence's energized line wires and the soil or grounded wires

direct register. Description of a gait in which the rear foot lands on top of the print made by the front foot on the same side

diurnal. Most active during daylight hours

dog fox. A male fox

electroplastic netting. Electrified plastic netting used for fencing

energizer. The unit that powers an electric fence

ermine. A short-tailed weasel in its white winter coat. See stoat

Falconiformes. Members of a taxonomic order of diurnal birds of prey known as falcons

fovea. A focal point at the back of the eyeball that functions as an image enlarger

gait. The order and speed with which an animal's feet touch the ground as it travels

grounding. Providing an electric fence's current an easy path through the soil; also called earthing

hood. The crown and nape feathers on a bird

hop. An elongated gait in which all four feet leave the ground twice in each cycle, resulting in the hind foot prints landing ahead of the front prints; also called gallop

impedance. The measure of how much resistance is present to restrict energy flow in an electric fence system

indirect register. Description of a gait in which the rear foot falls just behind, or overlaps, the print made by the front foot on the same side

irruption. A sporadic mass migration

joule. A unit of energy representing one amp of current through a one-volt force for one second

latrine. A spot where a single animal (like a dog or cat) periodically returns to deposit scat, or several animals in a family (like raccoons) deposit scat over a period of time

leak. A diversion of energy from an electric fence caused by such things as weeds or faulty insulators

matutinal. Most active at dawn

mesocarnivore. An animal for which meat is at least half of its diet

mobbing. A prey species' technique of cooperatively harassing or attacking a potential predator

monocular vision. Using the eyes separately, so that each looks at a different object at the same time

negative space. In an animal's track, the area between the toes and the paw pad

nocturnal. Most active after dark

omnivore. An animal for which meat makes up less than 30 percent of its diet

passerine. A large perching bird

pellet. A cylindrical or round mass of undigestible feed coughed up by a wild bird; also called cough pellet

polecat. Alternative name for a weasel or a skunk

polywire. Electric fence material consisting of several polyethylene strands twisted together with stainless-steel filaments for conductivity; also called electroplastic twine

raptor. Collective designation for eagles, falcons, hawks, owls, and vultures based on similarities in their appearance and behavior

righting reflex. The ability of cats to rotate their body so that they land on all four feet when they fall

scare wires. Electrified line wires used to keep animals from rubbing or leaning against a nonelectric fence, or from digging or burrowing under it; also called offset wires

scat. Excrement, aka poop

scavengers. Birds (and other animals) such as vultures and condors that primarily eat carrion

scrape. Marks associated with scat resulting from claw scratching by an animal such as a cat, dog, or bear

segmented. With reference to scat, sectioned at right angles into smaller pieces

short. A short circuit in an electric fence; also called a voltage leak

small-mesh field fence. A type of wire livestock fence with small openings; also called yard fencing, lawn and garden fencing, small stock fencing, goat and sheep fencing, kennel fencing, or poultry fencing

sole walker. A five-toed animal that walks flat-footed (plantigrade)

solitary. Living and foraging or hunting alone

stoat. A short-tailed weasel in its brown winter coat. See ermine

straddle. The width of an animal's trail pattern, from the outside of the left track to the outside of the right track

stride. The distance between a spot on an animal's track and the same spot on the next track made by the same foot

Strigiformes. Members of a taxonomic order of mostly nocturnal birds of prey known as owls

talon. A claw, usually in reference to a bird of prey

tension fence. A fence constructed with tightly tensioned wire of high-tensile strength

territory. The geographic area within which a single animal, family, or clan lives and hunts; also called home range

toe walker. An animal that walks on its toes (digitigrade); also called ball walker

tonic immobility. A natural state of paralysis used defensively by an animal to cause a predator to lose interest; also known as playing dead or playing 'possum

track. An animal's footprint; also, the act of following an animal's footprints

trail. A series of tracks reflecting an animal's movement

trail pattern. A series of tracks reflecting an animal's gait; also called track pattern

trap smart. Describes an animal that once was trapped but then escaped and thereafter avoids traps

trot. A gait for which diagonally opposite legs move together, resulting in a long stride and narrow straddle

true weasel. A member of the weasel family belonging to the genus *Mustela*

uric acid. The white part of bird and reptile droppings; the equivalent of human urine

vespertine. Most active in the evening, between sundown and nightfall

vixen. A female fox

volt. The amount of force that moves an electric current

vulture. A raptor of the Cathartidae family

waddle. To walk by moving both feet on one side of the body before moving both feet on the other side

wake. A group of vultures dining on a carcass

walk. A slow gait in which the legs move independently and at least one foot has contact with the ground at all times

welded wire. A type of fencing in which the horizontal and vertical wires are welded together wherever they meet

woven wire. A type of fencing in which the horizontal and vertical wires are woven, knotted, or hinged together wherever they meet

yoke-toed. A feature characteristic of owls, in which two toes point forward and the other two point back

Resources

ADDITIONAL READING

For in-depth information on wildlife tracks and signs, you can't beat the following two field guides:

Elbroch, Mark, with Eleanor Marks. *Bird Tracks & Sign: A Guide to North American Species*. Stackpole Books, 2001.
An illustrated guide to identifying bird families and species by the many signs they leave

Elbroch, Mark, and Casey McFarland. *Mammal Tracks & Sign: A Guide to North American Species, Second Edition*. Stackpole Books, 2019.
Range maps and illustrated descriptions of tracks, trails, scat, and other signs left by mammals

OTHER USEFUL BOOKS

Dohner, Janet Vorwald. *The Encyclopedia of Animal Predators*. Storey Publishing, 2017.
Profiles 50 mammals, birds, and reptiles that prey on poultry, pets, and livestock

Halfpenny, James C. *Scats and Tracks of North America*. The Globe Pequot Press, 2008.
Compact yet detailed guide to animal scat and tracks, organized by family groups

ORGANIZATIONS

For information on rules and regulations regarding dealing with a poultry predator, or if you are simply at your wit's end over a persistent predator, the following agencies may be of assistance:

United States Fish and Wildlife Service (www.fws.gov) offers information on federal wildlife protection regulations and permits.

Wildlife Service, United States Department of Agriculture Animal and Plant Health Inspection Service (www.aphis.usda.gov), provides contact information for agencies in your state through the "wildlife" links.

Association of Fish and Wildlife Agencies (www.fishwildlife.org) helps you find your local state wildlife agency by following the "members" link.

ONLINE RESOURCES

All About Birds
www.allaboutbirds.org
Descriptions, photos, and audio recordings that help identify a specific raptor, corvid, or gull

Animal Tracks Den
http://bear-tracker.com
Excellent information on tracks and scat of various species, with lots of photos

Bear Smart
http://bearsmartdurango.org
Addresses the behaviors of people that result in human-bear conflict, and offers solutions to living responsibly in bear country

Raccoon Latrines: Identification and Clean-up
CENTERS FOR DISEASE CONTROL
www.cdc.gov/parasites/baylisascaris/resources/raccoonlatrines.pdf
Describes the hazards of handling raccoon scat and the proper procedure for its removal

Cougar Network
www.cougarnet.org
A research group that tracks confirmed cougar sightings and offers free PDFs "Puma Identification Guide" and "Puma Field Guide"

Dig Defence
http://digdefence.com
In-ground fence panels that extend the protection of any fence or enclosed area to exclude digging predators

How to Cast Tracks
ORANGE COUNTY TRACKERS
http://octrackers.com/TRACKCASTING.htm
Simple directions for making permanent casts of animal tracks

iTrack Wildlife Animal Tracks App
www.naturetracking.com/itrack-wildlife
A comprehensive digital offline field guide to animal tracks

ONLINE RESOURCES, continued

Interagency Grizzly Bear Committee

http://igbconline.org

Lists certified bear-resistant products and offers tips on staying safe in bear country

Kencove Farm Fence Supplies

http://kencove.com

Complete source for constructing all types of electric fence, with lots of informative articles and blogs

Mountain Lion Foundation

www.mountainlion.org

Details on mountain lion biology and behavior, and suitable methods for protecting people, pets, and livestock

Nature Tracking

www.naturetracking.com

Online guides for identifying mammal, bird, reptile, and amphibian tracks

Nuisance Wildlife Trapping

WASHINGTON DEPARTMENT OF FISH & WILDLIFE

http://wdfw.wa.gov/living/nuisance/trapping.html

Tips on how and when (and when not) to trap nuisance wildlife and what to do with a trapped animal

PredatorPee

www.predatorpee.com

Offers authentic urine of various predators for pretend-marking a territory to deter unwanted wildlife from entering

Prevent Rodent Infestations

CENTERS FOR DISEASE CONTROL

www.cdc.gov/rodents/prevent_infestations/index.html

How to keep out rodents and safely clean up existing rodent nests and droppings

Professional Wildlife Removal

www.wildlife-removal.com

Nationwide directory of experts specializing in wildlife control

RollGuard

www.coyoteroller.com

Manufacturer of a device that prevent predators from getting the foothold they need to climb over a fence.

TrailCamPro

www.trailcampro.com

Descriptions, reviews, and sample photos comparing various motion-activated trail camera brands and models

The Wolverine Foundation

http://wolverinefoundation.org

How to identify a wolverine and report a sighting

For more about the lives of many predators discussed in this book, follow this link to a supplementary resource on the Storey Publishing website:
https://www.storey.com/poultry-predators/

Sherlock Holmes, Detective

The quotes from Sherlock Holmes can be found in the following books by Sir Arthur Conan Doyle:

1. **Scene of the Crime**
   "There is nothing like first-hand evidence."
   —*A Study in Scarlet*

2. **Who Coulda (or Couldn'ta) Dunnit?**
   "Eliminate all other factors, and the one which remains must be the truth."
   —*The Sign of Four*

3. **Sleuthing for Clues**
   "In an investigation, the little things are infinitely the most important."
   —*The Art of Detection*

4. **Foiling the Perps**
   "It is better to learn wisdom late than never to learn it at all." —*The Man with the Twisted Lip*

5. **Fence Defenses**
   "It is simpler to deal direct."
   —*The Adventure of the Sussex Vampire*

6. **The Poultry Perspective**
   "You'll get results by always putting yourself in the other fellow's place and thinking about what you would do yourself."
   —*The Adventure of the Retired Colourman*

Index

Page numbers in *italics* indicate images, tables, and charts.

Metric Conversions

1 inch = 2.5 cm
1 foot = 0.3 m
1 mile = 1.6 km
1 acre = 0.4 ha
1 ounce (liquid) = 30 ml
1 ounce (weight) = 28 g
1 pound = 4.5 kg

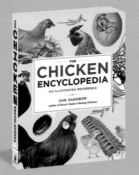

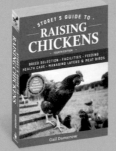